QUENCHING AND CARBURISING

QUENCHING AND CARBURISING

Proceedings of the Third International Seminar of the
International Federation for Heat Treatment
and Surface Engineering

Held in conjunction with the Annual Conference of
The Institute of Metals and Materials Australasia
(The Materials Society of IEAust)

THE INSTITUTE OF MATERIALS

Book 566
Published 1993 by
The Institute of Materials
1 Carlton House Terrace
London SW1Y 5DB

© The Institute of Materials 1993

Produced under the auspices of
The International Federation for Heat Treatment
and Surface Engineering

British Library Cataloguing in Publication Data

Quenching and Carburising: Proceedings of
the 3rd International Seminar of the
International Federation for Heat
Treatment (Melbourne, 1991)
620.1121

ISBN 0-901716-51-0

Typeset in Great Britain by
Alden Multimedia Ltd, Northampton

Printed in Great Britain by
The Alden Press, Osney Mead, Oxford

Contents

Introduction P.D. HODGSON	vii
1 State of the art in quenching B. LISCIC	1
2 Measurement and evaluation of the quenching power of quenching media for hardening J. BODIN AND S. SEGERBERG	33
3 Quench severity effects on the properties of selected steel alloys CHARLES E. BATES, GEORGE E. TOTTEN AND KIMBERLEY B. ORSZAK	55
4 Use and disposal of quenching media. Recent developments with respect to environmental regulations E.H. BURGDORF	71
5 Use of fluidised beds for quenching in the heat treatment field RAY W. REYNOLDSON	85
6 Gas quenching with helium in vacuum furnaces BENOIT LHOTE AND OLIVIER DELCOURT	107
7 Residual stress in quenched spheres D.W. BORLAND AND B.-A. HUGAAS	119
8 Computer simulation of residual stresses during quenching A.K. HELLIER, M.B. MCGIRR, S.H. ALGER AND M. STEFULJI	127
9 A mathematical model to simulate the thermomechanical processing of steel P.D. HODGSON, K.M. BROWNE, D.C. COLLINSON, T.T. PHAM AND R.K. GIBBS	139
10 Measurement and characterisation of air-mist nozzles for spray quenching heat transfer M.S. JENKINS, S.R. STORY AND R.H. DAVIES	161
11 Investigation of quenching conditions and heat transfer in the laboratory and in industry S. SEGERBERG AND J. BODIN	177

12	The design and performance of a laboratory spray cooling unit to simulate in-line heat treatment of steel R.E. GLOSS, R.K. GIBBS AND P.D. HODGSON	189
13	Microstructure, residual stresses and fatigue of carburised steels G. KRAUSS	205
14	Fundamentals of carburising and toughness of carburised components J. GROSCH	227
15	Martempering and austempering of steel and cast iron GEORG WAHL	251
16	Property prediction of quenched and case hardened steels using a PC T. RETI, M. GERGELY AND C.C. SZILVASSY	267
17	Effect of carburising on mechanical properties of steels NORIO KANETAKE	281
18	Heat treatment with industrial gases – Linde Carbocat and Carboquick processes ABDUL KAWSER	293
	Index	305

Introduction

Quenching and Carburising are two of the most basic and widely practised steel heat treatment processes. Each allows the base properties and performance of the steel to be significantly enhanced, such that a relatively inexpensive and simple starting material can be used for a wide range of demanding applications. Nevertheless, the technological developments within those two processes are often ignored in favour of 'high tech' surface treatments.

The aim of *Quenching and Carburising* was to review the recent advancements that have been made in these fields. The conference was the third biennial seminar of the International Federation for Heat Treatment and Surface Engineering (IFHT) and aptly formed the centrepiece of the Institute of Metals and Materials Australasia's (IMMA) annual conference, *Materials Processing and Performance*.

The sessions devoted to quenching covered a range of topics from the measurement of quench intensity, through to applications within the heat treatment industry and the steel processing industry. The current status of the application and science of quenching was reviewed by Professor Liščic from the University of Zagreb. Contributed papers covered developments in the use of fluidised bed furnaces, helium quenching in vacuum furnaces and the environmental issues associated with the disposal of quenching media.

As with most industrial processes there is now a strong need for accurate models of the quenching process, and a number of examples were given: quenching of high carbon steels, where the role of residual stress is important, air mist cooling during continuous casting, spray cooling in heat treatment and after hot rolling and gas quenching with helium. One important feature of such modelling is the generation of accurate data in the laboratory, and then incorporating differences which exist between controlled laboratory conditions and full-scale industrial plants. It is also apparent there there is now a degree of commonality between conventional heat treatment operations utilising controlled cooling and modern *inline* thermo-mechanical processing routes where the properties are generated during the shaping of the final product.

Carburising currently receives less exposure in the recent technical literature than quenching; largely because it is now an established and widely practiced technique for surface engineering. However, there are still a number of issues which need to be addressed, and these were covered in the two review papers by Professors G. Krauss (Colorado School of Mines) and J. Grosch (Technical University of Berlin), with particular emphasis on the microstructures and toughness of the carburised steels. The contributed papers cover modelling, mechanical properties and industrial carburising processes.

Overall the conference demonstrated that there is a great deal of knowledge regarding many aspects of *Quenching and Carburising*, and that the challenge is to apply this knowledge to control the process, and reduce the product variability. Steel is now a highly engineered material, which is still developing improved properties at lower cost to the user. To this end it is hoped that the material contained within this volume will enhance the application of these engineering processes throughout the heat treatment and thermomechanical processing industries.

The success of the conference was largely due to a small group of highly enthusiastic hardworking people, most notably Bruce Hinton (DSTO-ARL), Bill Sinclair (BHP Research) and the staff of the IMMA office, particularly Margaret Kirk and Angela Krepcik. Great support was also provided by members of the IFHT, and particular thanks is expressed to Professor T. Bell, Professor B. Liščić and Dr J. Naylor for their assistance during the conference organisation.

Finally, I would like to thank the authors, many of whom ventured down-under for the first time, who provided us all with the opportunity to learn, discuss and benefit from their experiences in heat treatment research and industrial application. They have helped create a volume which, it is hoped, will serve for many years as an important reference for those either working in the heat treatment industry, or teaching these subjects.

Peter D. Hodgson
BHP Research – Melbourne Laboratories
Mulgrave, Victoria, Australia.
Conference Chairman

1
State of the Art in Quenching

B. LIŠČIĆ

Faculty of Mechanical Engineering and Naval Architecture, University of Zagreb

ABSTRACT

A survey of actual methods (based on direct temperature measurement) for quenching intensity evaluation in laboratory conditions as well as in workshop practice is given, and commented, especially in relation to possible prediction of metallurgical transformations and hardness distribution on the cross-section of real workpieces.

Apart from the widely used immersion quenching technique, the fundamental and latest developments of other quenching techniques such as spray quenching, gas quenching in vacuum furnaces and hot salt bath quenching are briefly described. Information about 'intensive quenching' is given, and the prospects of further developments in the field of quenching outlined.

1 INTRODUCTION

Quenching has a historical background centuries old, and has always played an important role in the manufacture of metallic products. Certainly, the technology of quenching has changed as well as the kind of workpieces being quenched: from axes and swords to gears and automotive components and, as the latest development, to a gas turbine blade that has been quenched in a hot isostatic pressing (HIP) quenching unit in argon under 2000 bars of pressure.[1] Even more important is the fact that simultaneously quenching has changed from an empirical skill to a scientifically founded and controlled process which now belongs to the area of 'intelligent processing' of materials.[2] This is the goal to which we are striving, and hopefully by the beginning of a new millenium we will be much closer to it than we are today, because in present workshop practice we still have to rely on practical experience.

For a long period in the recent history of quenching, besides water, oil was the main quenching medium. Correspondingly, manufacturers and suppliers of quenching oils have played the main role in the development of quenching technology and related testing methods and specifications.

Nowadays the assortment of quenchants in use is much broader. There are at least seven groups of quenching media having different chemical and physical

2 Quenching and Carburising

properties and, of course, different quenching intensities. They are (in descending order of quenching intensities):

- Water solutions of inorganic substances.
- Plain water.
- Polymer solutions in water.
- Quenching oils of different grades.
- Hot salt baths.
- Fluidised beds.
- Pressurised and circulated gases (N_2, He, Ar).

In spray quenching techniques water, water–oil emulsions, polymer solutions in water or water–air mixtures are used.

One has to take into account that with each of above mentioned groups of quenchants very many different quenching intensities can be realised. On the one hand this is due to different substances and their concentration, and to different quenching parameters (bath temperature, agitation rate, pressure) on the other. This situation calls for a universally applicable method for testing and evaluation of quenching intensity and makes the selection of optimum quenchant and quenching conditions both from the technological and economical point of view an important consideration.

Following modern production concepts such as 'just in time manufacturing' and 'total quality control', when reproducibility of results becomes necessary, heat treatment processes and relevant equipment have become fully automated and CNC controlled. Of all heat treatment operations quenching (especially immersion quenching) is still the least controlled. Even in a fully automated and CNC controlled sealed batch furnace, where temperature–time cycles as well as the atmosphere can be automatically changed according to the requirements at the time, quenching intensity depends on the quenchant and quenching conditions selected and nothing much can be done *during* quenching to change the quenching intensity in order to obtain optimum structures, hardness and residual stresses. This is unfortunately so, because with immersion quenching the only quenching parameter which can be changed in the course of quenching is the rate of agitation of the quenchant. This possibility is also limited to workpieces of bigger cross-sections (where the quenching itself lasts longer), and calls for an adequate sensor which will automatically change the speed of the agitation pump exactly at the moment when a certain temperature has been reached at the relevant point in the cross-section of the workpieces treated. The best possible conditions for a fully automated CNC control of the quenching intensity exist with water–air spray quenching.[3]

In order to predict the microstructure, the distribution of hardness (strength) and residual stresses after quenching as well as to optimise the hardening process, computer modelling is performed. The benefits of such optimised production process are as follows:[2]

– Reducing scrap or rework rate.

- Reducing trial and error in developing process parameters for new parts.
- Allowing tighter tolerances to be achieved via better control of the heat-treating process.

The most critical information relating to the hardening process is the rate of heat transfer from the workpiece to the quenching medium.

2 IMMERSION QUENCHING

Although quenching techniques, especially in recent decades, have become more and more diversified, immersion quenching still constitutes the largest part of all quenching processes.

Immersion quenching is mostly performed in evaporable quenchants like oils or water solutions, exhibiting the three known stages of cooling, i.e. vapour blanket stage, boiling stage and convection stage. It is a non-stationary heat transfer process that can be described using the following equation, the left side of which being Fourier's differential equation giving the heat flow from the interior to the surface of the workpiece, while the right side is Newton's equation for heat transfer from the workpiece surface to the ambience:

$$q = -\lambda \frac{\partial T}{\partial x_N} = \alpha (T_N - T_{amb}) \quad (1)$$

where:

q = heat flux density, W m^{-2}
λ = thermal conductivity within the workpiece, W/mK,
$\frac{\partial T}{\partial x_N}$ = temperature gradient normal to the surface of the workpiece, K/m,
α = interfacial heat transfer coefficient, W m^{-2} K,
T_N = surface temperature of the workpiece, K,
T_{amb} = ambient temperature, i.e. temperature of the quenchant, K.

In the case of quenching real parts there are many factors which influence this heat transfer/metallurgical transformation process:

(a) Factors depending on the workpiece itself:
   alloy grade (transformation characteristics);
   mass of the workpiece (size of the critical cross-section);
   geometry (volume/surface ratio);
   surface roughness and condition;
   load arrangement and density.*
(b) Quenchant characteristics:
   density (molecular weight);
   viscosity;
   specific heat;
   thermal conductivity of the fluid;
   boiling temperature;

Leidenfrost-temperature (transition temperature between the vapour blanket and the boiling stage);
wetting property of the liquid.

(c) Factors depending on quenching facility:
Bath temperature;*
Agitation rate;*
Flow direction;
Concentration of the solution (if applicable).*

Of all these influencing factors listed, only a few can be changed in the heat treatment shop. They are marked with a * and are usually called technological parameters of quenching.

Taking into account the simultaneous influence of so many factors during the quenching process, one logically wants to know what 'cooling power' or 'quenching intensity' will result in each specific case. There are actually three different reasons why we want to have as precise and complete information of the quenching intensity, as possible, namely:

I. Checking the specific characteristics of the quenchant (e.g. for mutual comparison, monitoring of deterioration or development of new grades of quenchants).
II. Selection of optimum quenchant and technological quenching parameters for the specified alloy and workpieces.
III. Computer modelling of resulting properties.

On the other hand many methods exist for measuring and evaluating quenching intensity and one has to know which method should be applied in the cases of I, II or III.

The IFHT Committee on Scientific and Technological Aspects of Quenching, established in 1978, has, from its very beginning, distinguished between a *laboratory* method for testing of the cooling power of a quenching medium, and a *practical* method for the measurement of the quenching intensity in the quench tank, assuming that laboratory testing is carried out in a small quantity of quenching medium without agitation, while practical testing involves all relevant technological parameters.

2.1 EVALUATION OF THE QUENCHING INTENSITY BY COOLING CURVES

Of all methods for testing quenching intensity, those measuring the temperature as a function of time (cooling curve) at a specified point within the test specimen are mostly used. It seems also to be generally accepted to use test specimens of cylindrical shape, but unfortunately there is no general agreement as yet on specimen size, material and position of the thermocouple tip. In order to standardize world-wide the *laboratory* method for testing of industrial quenching oils the IFHT Committee on Scientific and Technological Aspects of Quenching

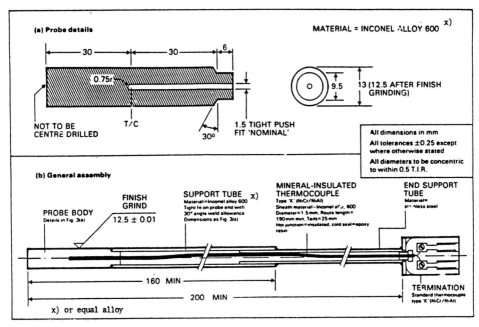

Fig. 1. Thermal probe for laboratory test for quenching oils.

has elaborated the Draft International Standard ISO/DIS 9950, called: 'Industrial Quenching Oils – Determination of Cooling Characteristics – Laboratory Test Method'. This standard method is based on a method drawn up by a working party of the Wolfson Heat Treatment Centre, the University of Aston in Birmingham (UK). It uses a 12.5 mm diameter × 60 mm cylindrical specimen made of Inconel alloy 600 with the thermocouple tip located at its geometric centre. Figure 1 shows the assembly of this thermal probe. As a result of each test, using this method, a temperature/time and (by differentiation of it) a temperature/cooling rate plot is obtained (Fig. 2).

It is to be hoped that the ISO Technical Committee TC-28 will soon approve this international standard for testing quenching oils. A similar draft of international standard for laboratory testing of polymer solutions is in preparation.

From temperature vs. time and cooling rate vs. temperature plots shown in Fig. 2, some characteristic features can be obtained, e.g. the transition temperature between vapour blanket stage and boiling stage (T_L); the maximum cooling rate (CR_{max}) as well as the cooling rate at 300°C (CR_{300}), indicating the tendency for cracking and distortion. Nevertheless, the main problem still remains: which method should be employed for cooling curves interpretation? While checking oils by the laboratory test there is no agitation of the quenchant. But by applying the same method for testing polymer solutions, agitation and bath temperature (technological parameters) must be involved, otherwise no reproducible results can be expected.

6 Quenching and Carburising

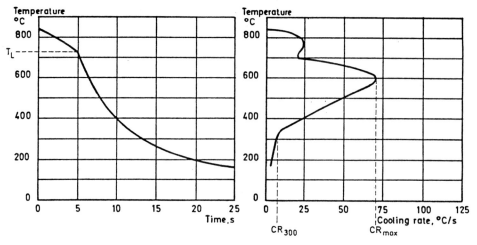

Fig. 2. Typical temperature/time and temperature/cooling rate plots from test probe cooled in a quenching oil.

First, visual characterisation and comparison of cooling curves was used to correlate them with metallurgical properties, i.e. hardness distribution obtained after quenching. Figs. 3(a) and 3(b), taken from Segerberg[4] show a case of such comparison where a polymer solution of 15% concentration (Aqua-quench 251) at 30°C and 50°C was compared to quenching oil (Gulf U) of 70°C, using the hardness distribution measured on specimens of 16 and 25 mm diameter made of E_n-43 B steel grade. After quenching in Aquaquench 251 at 30°C a higher hardness was obtained rather than after quenching in oil (being in accordance with the cooling rate curves in Fig. 3a); quenching in Aquaquench 251 of the same concentration at 50°C resulted in a lower hardness than that after quenching in oil (Fig. 3b), which is not in accordance with the cooling rate curves shown.

Thelning[5] has shown that in correlating cooling curves obtained by the laboratory testing method with metallurgical properties, the use of single characteristic features was useful but not truly representative of the properties obtained. Later, other methods were suggested for cooling curves interpretation, as follows:

- By using linear regression analysis, N.A. Hilder,[6] quantitatively related variation in polymer concentration, bath temperature and agitation rate to the cooling curve behaviour of polymer-solution quenchants by the following formulae:

$$CR_{max} = 244.7 - 4.3\ C - 1.7\ T + 47.2\ V \qquad (2)$$
$$CR_{300} = 83 - 1.68\ C - 0.36\ T + 14.1\ V \qquad (3)$$

where:
C = polymer concentration in volume percent,
T = bath temperature in °C,

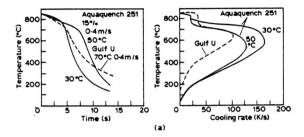

Fig. 3(a). Cooling curves and cooling rate curves of a polymer solution (Aquaquench 251) of 15% concentration 0.4 m/s agitation rate and an oil (Gulf U) of 70°C and 0.4 m/s agitation rate.

V = agitation rate in m/s.

- An alternative procedure[7] for characterising cooling curve shape utilizes a mathematical equation to describe the entire curve, in order to study the characteristic features of the curve in relation to quenchant properties. The cooling curve equation is a non-linear function where the temperature depends on seven parameters which can be estimated for any cooling curve using a computerised, iterative non-linear least squares algorithm.
- Thelning[5] suggested quantifying the cooling capacity of a quenchant bath by calculating the area under the cooling rate curve between two predetermined temperatures. Other authors[8] have tried to relate the values for maximum cooling rate (CR_{max}) and cooling rate at 232°C (CR_{232}) with the area under the cooling rate curves for various quenching media, but they did not find a logical correlation.

Experience so far using the laboratory testing method with a small diameter (12.5 mm) and the mass of the probe shows that this method and relevant cooling curves analysis should be used for checking the specific characteristics of the quenchant (reason I), i.e. for:

– determination of the vapour blanket stage

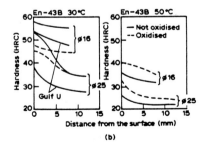

Fig. 3(b). Hardness distribution curves on the cross-section of 16 and 25 mm diameter specimens made of E_n-43 B, after quenching under conditions from Fig. 3(a).

- evaluation of the maximum cooling rate and the cooling rate within the martensite transformation range
- comparison of quenchants from different sources
- monitoring of the contamination, thermal degradation, oxidation or additive depletion of the quenchant
- information on the influence (for polymer solutions) of polymer concentration, bath temperature and agitation rate.

The cooling curves obtained by the laboratory testing method should not be correlated with metallurgical results (hardness distribution) on the cross-section of real workpieces, because of the following reasons:

- The small probe of 12.5 mm diameter used in laboratory testing cools down much quicker than real parts (e.g. in still mineral oil from 850°C to 200°C in about 20 s while the centreline of a 50 mm diameter cylinder needs about 700 s). The small diameter probe, accordingly, cannot record the heat transfer phenomena which are taking place in a real workpiece in the period after 20 seconds from immersion.
- Neither the large influence of the mass (or size) of the workpiece nor the influence of real quenching conditions (agitation rate and flow direction, loading arrangement and density), can be ignored.

Recently, Bates and Totten[9] have taken another approach to cooling curve interpretation. They have used cylindrical probes of 13, 25, 38 and 50 mm diameter made of AISI 304 stainless steel with the thermocouple tip located in the geometrical centre of the probes. After recording the centreline cooling curve, they have calculated the 'quench factor' and the value of the heat transfer (film) coefficient. The heat transfer coefficient provides the measure of the effective rate of heat removal from the workpiece. Therefore, this approach, which takes into consideration also the transformation characteristics of the steel concerned (through the 'quench factor' that will be described later), is much more realistic when relating cooling curves to the hardness of the quenched parts. Only two objections can be raised to this approach:

– Positions of points in which temperatures have been measured are too far from the surface of the probe (especially for bigger diameters), so the substantial damping effect must be taken into account.
– The heat transfer coefficient during a quenching process changes its value for more than one order of magnitude and is therefore difficult to present it by a sole number only.

2.2 EVALUATION OF THE QUENCHING INTENSITY BY THE HEAT FLUX DENSITY

Recently, several authors have recognised that the heat flux density and heat transfer coefficient represent the right approach to the evaluation of the quenching

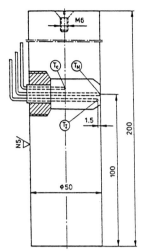

Fig. 4. Scheme of the LIŠČIĆ-NANMAC probe for measurement of the temperature gradient on the surface.

intensity especially in workshop conditions when probes and specimens of bigger mass are used.

Based on the research work done at the University of Zagreb the *temperature gradient method* has been developed as a new method for measuring, recording and evaluating the quenching intensity in workshop conditions.

It is universally applicable to all quenchants, quenching conditions and quenching techniques. It is based on the known physical rule expressed by equation (1), i.e. that the heat flux density at the surface of a body is directly proportional to the temperature gradient at the surface multiplied by the thermal conductivity of the body material.[10,11]

The essential feature of the new method lies in measuring and recording the temperature gradient at the surface of a special cylindrical probe during the entire quenching process. The temperature on the surface itself (measured by fast-response thermocouple for transient thermal measurements, US Pat. No. 2,829.185) and the temperature at a point just underneath (1.5 mm below) the surface, are measured in order to determine the temperature gradient at the surface and to calculate the heat flux density vs. time and the heat flux density vs. surface temperature, respectively.

Figure 4 shows a 50 mm diameter × 200 mm long cylindrical probe developed in cooperation with the NANMAC Corp., Framingham Centre, Massachusetts (USA). The probe is made of AISI 304 stainless steel, having a gland nut with three thermocouples assembled on the same radius at mid-length. The third thermocouple in the centre of the cross-section measures the centreline temperature and enables one to calculate the temperature difference core-surface vs. time which is important for calculation of thermal stresses during quenching.

The specific features of the probe are:

10 Quenching and Carburising

- The response time of the surface thermocouple is 10^{-5} seconds, so the fastest temperature changes can be recorded.
- The intermediate thermocouple can be positioned with an accuracy of ± 0.025 mm.
- The surface condition of the probe can be maintained by polishing the sensing tip before each measurement.
- The body of the probe, made of an austenitic stainless steel, does not change in structure during the heating-quenching process, nor does it evolve or absorb heat because of phase changes.
- The size of the probe and its mass ensure a sufficient heat capacity and a symmetrical radial heat flow in the cross-sectional plane where the thermocouples are located.
- There is no dependence of the heat transfer coefficient during boiling stage on the probe diameter, because such a dependence, according to Kobasko[12], exists only for diameters of less than 50 mm.

For each practical measurement, the probe is heated up to 850°C and transferred quickly (in about 3 s) to the quenching bath and immersed. The probe is connected to a microcomputer by an interface having three A/D converters and three amplifiers. The temperatures are recorded for 500 seconds after immersion. A total of 960 data points is recorded during each quench from every thermocouple.

Adequate software is developed to store the temperature vs. time data from all thermocouples and to calculate and display graphically the following functions:

$q = f(t) =$ heat flux density as a function of time W m^{-2}
$q = f(T_N) =$ heat flux density as a function of surface temperature W m^{-2}
$\Delta T = T_C - T_N =$ temperature difference between core and surface as a function of time.

Figure 5 shows these functions for quenching in mineral oil at 20°C without agitation and Fig. 6 for quenching in a polymer solution (UCON-E) of 25% concentration, 40°C bath temperature and 0.8 m/s agitation rate.

By taking the measured surface temperatures vs. time as input values, a computer program was developed to calculate and display graphically the heat transfer coefficient on the probe's surface as a function of time or as a function of surface temperature. Figures 7a and b show the calculated heat transfer coefficient for quenching in the mineral oil at 20°C without agitation, and Figs. 8a and b, for quenching the probe in the UCON-E polymer solution of 25% concentration, 40°C bath temperature and 0.8 m/s agitation rate. Once the function of the heat transfer coefficient is known for a specified quenching process, by solving the direct problem of heat conduction, temperature vs time (cooling) curves can be calculated at every arbitrary point on the cross section.

On comparison of the diagrams in Fig. 5(b) and 6(b) it is evident that with oil quenching the heat flux density reaches its maximum of 2.7 MW m^{-2} in the 16th second after immersion, while with the above described polymer solution quenching it reaches its maximum of 2.2 MW m^{-2} not earlier than 75 seconds after

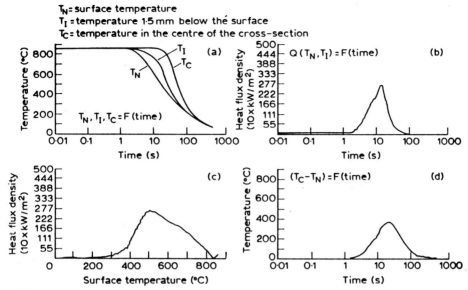

Fig. 5. (a) Temperature vs. time; (b) heat flux density vs. time; (c) heat flux density vs. surface temperature; and (d) temperature difference core-surface vs. time when quenching the probe in a mineral oil at 20°C, without agitation.

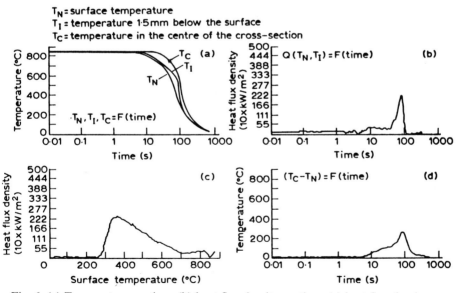

Fig. 6. (a) Temperature vs. time; (b) heat flux density vs. time; (c) heat flux density vs. surface temperature; and (d) temperature difference core-surface vs. time when quenching the probe in UCON-E polymer solution of 25% concentration; bath temperature 40°C; agitation rate 0.8 m/s.

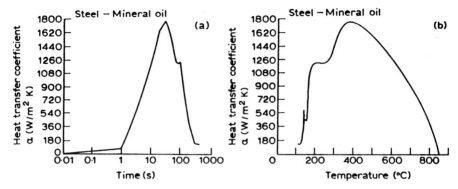

Fig. 7. (a) Calculated heat transfer coefficient vs. time and (b) vs. surface temperature when the probe was quenched in mineral oil at 20°C, without agitation.

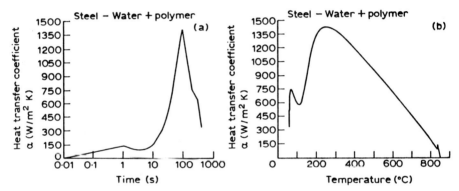

Fig. 8. (a) Calculated heat transfer coefficient vs. time and (b) vs. surface temperature when the probe was quenched in UCON-E polymer solution at 25% concentration; bath temperature 40°C; agitation rate 0.8 m/s.

immersion. Because the integral of the surface below the heat flux density curve during a certain period represents the quantity of heat extracted J m^{-2}, this is the main criterion in evaluating the real quenching intensity. According to this criterion, oil extracts more heat in shorter time than 25% polymer solution and should give higher hardness (or greater depth of hardening). Figure 9 shows measured hardness distributions on the cross-section of a 32 mm diameter specimen made of AISI-C1043 steel after quenching in water, oil and different concentrations of UCON-quenchant E (UQE). As can be seen, oil has actually given higher hardness values than UCON-quenchant E of 25% concentration.

Comparing Fig. 5c and Fig. 6c we can see that with oil quenching the maximum heat flux density occurred at 500°C surface temperature, while with the above described polymer solution the maximum heat flux density occurred at 350°C

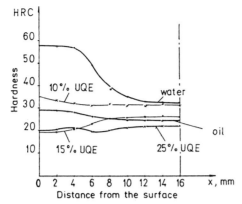

Fig. 9. Hardness distribution curves on the cross-section of the 32 mm diameter specimen made of AISI-C 1043 after quenching in different quenchants.

surface temperature, and at 300°C surface temperature the heat flux density was still remarkable (1.2 MW m^{-2}). This shows that in the low temperature region the polymer solution cools more quickly than oil, a fact that has to be taken into account in regard to martensite formation and residual stresses or risk of deformation and cracking, respectively.

Therefore, the *temperature gradient method* has the following advantages:

- It measures and records the real quenching intensity during the whole quenching process of actual workpieces.
- There is an unambiguous correlation between the criterion of quenching intensity and the hardness distribution (depth of hardening).
- It provides input data for calculating the course of actual heat transfer coefficient during the whole quenching process, and enables computer simulation of cooling curves at each point of the cross-section (reason III).

2.3 QUENCH FACTOR ANALYSIS

While previously described methods of quenching intensity evaluation do not consider the metallurgical transformation behaviour of the steel concerned, Bates,[13] building upon the work of Evancho and Staley, has demonstrated a general procedure for interrelating quenching variables and transformation kinetics of steel, i.e. the cooling curve and the relevant CCT diagram, providing a single number that indicates the extent to which a part can be hardened at various locations within the part.

The quench factor concept means that the hardening behaviour of steel during continuous cooling can be predicted by breaking a cooling curve into discrete temperature time increments and determining the ratio of the amount of time the steel was at each temperature, divided by the amount of time required to obtain a specific amount of transformation at that temperature. The sum of the incremental

14 Quenching and Carburising

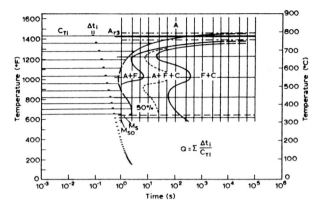

Fig. 10. Schematic illustration of the method for calculating a quench factor, according to Bates.[13]

quench factor values over the transformation range between A_{r3} and M_s is the quench factor Q. Quench factors can be calculated relatively easily from digitally recorded time-temperature curves (cooling curves) obtained in instrumented probes by using an equation given in Bates[13] which describes the C-curve (start of transformation in a CCT diagram) for the alloy of interest (C_T-function). In this calculation several constants, $K_1 \ldots K_5$, are related to the steel in question.

First, an incremental quench factor, q, for each time stop in the transformation range is calculated as:

$$q = \frac{\Delta t}{C_T} \qquad (4)$$

where:
q = incremental quench factor
Δt = time step used in data acquisition
C_T = the time for the start of transformation at the sample temperature.

The incremental quench factor values are summed over the transformation range between A_{r3} and M_s (see Fig. 10), to produce the cumulative quench factor, Q, according to the following equation:

$$Q = \Sigma q = \sum_{T=M_s}^{T=A_{r3}} \frac{\Delta t}{C_T} \qquad (5)$$

The quench factor reflects the heat extraction characteristics of the quenchant as a function of the agitation rate and bath temperature by virtue of agitation rate and temperature effects on the cooling curve shape. Q also reflects section thickness effects on the cooling curve and the transformation kinetics of the alloy, by taking into account the location of the C_T curve in time. Using the quench factor analysis procedure after quenching instrumented probes of different section sizes in different quenchants, it is possible to relate the hardness to the resulting cooling

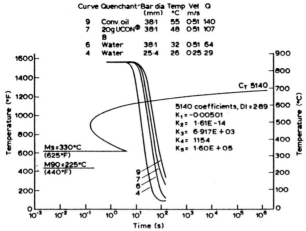

Fig. 11. Cooling curves plotted in log time superimposed on AISI 5140 C_T-curve, according to Bates.[13]

curve and to parts of the same shape and size made of an alloy with known C_T function, but always only for the point on the cross-section in which the temperature has been measured.

Continuous cooling curves plotted in log. time when quenching 38.1 mm and 25.4 mm instrumented probes in different quenchants and quenching conditions (see Fig. 11) are superimposed on the C_T curve for AISI-5140 steel. The curve numbers correspond to the data in Table 1. The coefficients that describe the C_T curve for AISI-5140 steel ($K_1 \ldots K_5$) have been obtained by the regression analysis of cooling curves and centreline hardness data of this steel (see Fig. 11). It can be seen from this figure that cooling curves that passed further into the C_T curve produced progressively higher quench factors. From Table 1 it is evident that higher quench factors in accordance with the crossed-section of the CCT diagram correspond to lower hardness values.

From the cooling curves recorded by instrumented probes cooling rates at 700°C and 260°C, interface heat transfer coefficients and quench factors have been calculated as can be seen from Table 1. The predicted hardness values in the centre of the specimen cross-section were calculated from the relevant quench factor, using the following equation:

$$P(\text{HRC}) = \text{HRC}_{\max} \cdot \exp(k_1 \cdot Q) \tag{6}$$

where:
$P(\text{HRC})$ = predicted hardness in HRC,
$\text{HRC}_{\max}$ = maximum attainable hardness in relevant alloy (HRC),
$\quad k_1 = l_n(0.995) = -0.00501$ (fraction untransformed austenite that defines the C_T curve),
Q = quench factor.

As can be seen from Table 1, the hardness values predicted from the quench factor

16 Quenching and Carburising

Table 1. Cooling rate and quench factor data on 5140 steel in several quenchants

Tape no. File no.	Quenchant	Probe diam (mm)	Austenite temp. (°C)	Bath temp. (°C)	Velocity (m/s)	Cooling Rate at 700C (C/s)	Cooling Rate at 260C (C/s)	Film coeff. (W/cm²·K)	Quench factor	Pred. hardness (Rc)	Actual hardness (Rc)
1	Water	12.7	843	48	0.25	162	151	1.53	9.2	53.5	53.1
2	Water	12.7	843	60	0.25	75	144	0.59	13.1	52.5	54.2
3	Water	12.7	843	71	0.25	42	132	0.29	21.5	50.3	51.4
4	Water	25.4	843	26	0.25	111	41	1.87	29.2	48.4	46.0
5	Water	25.4	843	60	0.25	30	37	0.71	35.0	47.0	45.5
6	Water	38.1	843	32	0.51	57	20	1.50	63.1	40.9	40.5
7	20% UCON®B	38.1	843	48	0.51	33	8	0.40	107.4	32.7	33.0
8	Fast oil	38.1	843	65	0.51	43	10	0.68	88.6	36.0	34.0
9	Conv. oil	38.1	843	65	0.51	32	7	0.29	139.6	27.8	29.5
10	Martemp oil	38.1	843	148	0.51	38	4	0.36	125.3	29.9	31.0

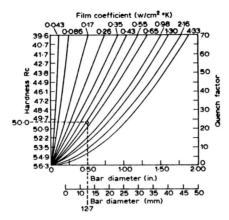

Fig. 12. Correlation of quench factor data, bar diameter and hardness for AISI-5140 steel, according to Bates.[13]

were reasonably close to the actual measured hardness with all quenchants and section sizes. The correlation coefficient between the predicted and measured hardness values was 0.975.

Besides the functional relationship that exists between as-quenched hardness and the quench factor, other relationships exist *for a specific alloy* between the interface heat transfer coefficient, section thickness and quench factor. Figure 12 illustrates the effect of bar diameter and interface heat transfer coefficient on the quench factor values and as-quenched hardness at the centreline of the AISI-5140 steel.

If, for example, in the centreline of a 12.7 mm bar the hardness of 50 HRC with the AISI-5140 steel is to be achieved (see Fig. 12), then a heat transfer coefficient between 0.55 and 0.65 W cm^{-2} K is to be realised, and adequate quenchant and quenching conditions selected. According to Table 1 (File no. 2) this could be achieved by quenching in water at 60°C and 0.25 m/s agitation rate (this gives film coefficient of 0.59 W cm^{-2} K).

The quench factor concept which deserves full attention and is one of the most realistic methods for practical selection of quenchants and/or quenching conditions when quenching a particular alloy (reason II) has, of course, in practice its weaknesses and shortcomings, e.g.

- The selection of specific quenching conditions based on the desired heat transfer coefficient value is possible only when a sufficient number of experiments has been performed previously with the relevant quenchant, otherwise the heat transfer coefficient value for a new quenchant and quenching conditions is not known.
- The influence of some quenching parameters, e.g. bath temperature on the heat transfer coefficient single value calculated in this way, is sometimes very large. If, for example, the above given specimen would be quenched in water at 48°C

18 Quenching and Carburising

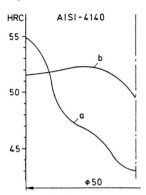

Fig. 13. Comparison of hardness distribution curves on 50 mm dia AISI 4140 steel after quenching in: (a) conventional oil of 20°C, without agitation; (b) 20% solution of UCON-E polymer, bath temperature 35°C, agitation rate 1 m/s.

(see Table 1) instead of water at 60°C – other quenching conditions equal – the heat transfer coefficient would be 2.6 times greater! (1.53 instead of 0.59 W cm^{-2} K). As we know, it is quite difficult, in practice, to find out the temperature of the quenchant surrounding the workpieces being quenched within an accuracy of ±10°C.
- By predicting the hardness at one point of the workpiece's cross-section, it is not possible to draw a conclusion about the course of the hardness distribution curve. Sometimes, it can be even misleading. The following example demonstrates this.

By comparing three tests of quenching in a 20% polyglycol solution (see Ref. 13, p. 41, Table 6, File no. 13, 14, 15) with quench factors 55.1 and 56.7, with three tests of quenching in conventional oil (see Ref. 13, p. 41, Table 5, File no. 31, 32, 33) giving quench factors: 65.6; 66.6; 65.3 – for the same bar diameter and same steel grade – the author concludes that: "These data indicate that the 20% polymer solution is equivalent or slightly superior to the conventional oil in its ability to harden the steel." But that such a conclusion can be misleading shows another example:

In a graduation thesis[14] carried out at our faculty in Zagreb hardness distribution curves have been measured on the cross-section of a 50 mm bar made of AISI-4140, after quenching in a conventional oil at 20°C, without agitation and in a 20% solution of UCON-E polymer, bath temperature 35°C, agitation rate 1 m/s. Figure 13 shows that in this case it would not be correct to apply this conclusion, because although the centreline hardness is 8.5 HRC higher for a polymer-solution quenched specimen, the surface hardness is 3.5 HRC higher for the specimen quenched in the conventional oil. While the oil quenched specimen shows a normal hardness distribution curve, the specimen quenched in the described polymer solution showed an inverse distribution curve. This example makes clear that prediction of hardness at one point of the cross-section only is not enough to conclude the hardness distribution, but computer aided modelling of the hardness values in at least five points (for cylindrical cross-sections: surface; 3/4 R; 1/2 R; 1/4 R and centre) is necessary. For such modelling (reason III), the *temperature gradient method*, with calculated course of heat transfer coefficient during whole quenching process and calculated cooling curves in all five mentioned points – superimposed on the relevant CCT diagram – provide the right answer.

To conclude this section on immersion quenching, being a very complicated process, two more phenomena which can influence the quenching process and prediction of hardness should be mentioned:

– One is the wetting process (wetting kinetics) that takes place during quenching along the specimen's surface.[15] This phenomenon has been studied by Tensi at the University of Munich (Germany).
– The other phenomenon, studied by Shimizu and Tamura,[16] deals with transformation kinetics of the steel concerned. When using CCT diagrams for hardness prediction and cooling curves of quenchants where cooling rates are *discontinuously* altered during cooling, the transformation behaviour of steel depends also on the incubation period, consumed before changing the cooling rate. Without taking into account this phenomenon, it is not possible to explain the inverse hardness distribution pattern that sometimes occurs.

3 SPRAY QUENCHING

The main difference between immersion quenching and spray quenching is that by using immersion quenching no corrective action (besides changing the agitation regime) can be applied throughout the duration of the cooling operation, while with spray quenching the heat flux density and the heat transfer coefficient during the cooling operation itself can be controlled.

The heat transfer from the metal body to the cooling fluid is, according to Jeschar *et al.*[17] mostly defined by the velocity distribution within the fluid which depends on the vapour blanket thickness. If the velocity distribution is caused not only by rising vapour (as with immersion quenching) but also by a coercive stream, the heat transferred to the fluid will be bigger. It is important that with spray cooling the heat transfer depends very much on the quantity of the sprayed water.

Figure 14 shows the scheme of the heat transfer in the vapour blanket stage, in the case where a heated metallic surface is being spray-cooled by water. The heat transfer coefficient to water (α_w) consists of an internal heat transfer resistance inside the water layer and a capacitive transport resistance for the sprayed amount of water:

$$\alpha_w = \left(\frac{1}{\alpha_{w,i}} + \frac{1}{m_s C_w} \right)^{-1} \tag{7}$$

where:

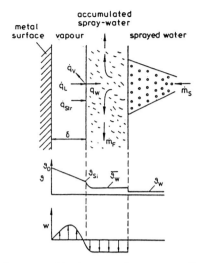

Fig. 14. Heat transfer in the vapour blanket stage for spray cooling, according to Jeschar et al.[17]. q_v = boiling heat flux density; q_L = conduction heat flux density; q_{str} = radiation heat flux density; q_w = heat flux density transferred to water; m_s = flux of sprayed water; m_F = flux of water flowing off; δ = thickness of the vapour blanket; ϑ_o = surface temperature; ϑ_{si} = boiling temperature; ϑ_w = water temperature; w = velocity of the fluid.

$\alpha_{w,i}$ = internal heat transfer coefficient of water, W m^{-2} K^{-1},
m_s = flux of sprayed water, kg m^{-2} s^{-1},
C_w = specific heat capacity of water, J kg^{-1} K^{-1}.

Figure 15 shows the total heat transfer coefficients (α_{tot}) vs. surface temperature of the workpiece (ϑ_o) for water spray-cooling, calculated from measurements. The parameter is the density of water striking kg m^{-2} min. It is evident that the values of α_{tot} can be much higher than in case of water immersion quenching, and that with higher density of water striking the heat transfer coefficient values are bigger. While in the temperature region of vapour blanket the heat transfer coefficient values do not depend on surface temperature (ϑ_o), their biggest changes are in the temperature region below the vapour blanket.

As is seen from Fig. 16, when water spraying is used to cool a metallic part from a high temperature of about 800°C, two successive cooling states are observed:

– First a non-wetting state, when water droplets do not touch the solid wall – this produces slow cooling, and
– Second, a subsequent wetting state, when droplets are spread over the solid wall, allowing high heat flux densities to be extracted.

Spray quenching can be realised by spraying water only, or by spraying a mixture of water and air. The pressures of water and air can be controlled separately, and the cooling rate varies with air and water pressure. By using a CNC control system this method of quenching single workpieces makes it possible to change continually the heat flux extracted, and to follow a predetermined cooling curve.

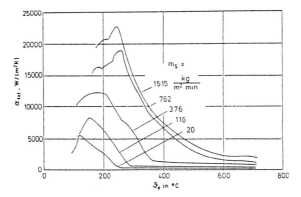

Fig. 15. Influence of the density of water striking on the heat transfer coefficient in case of water spray-cooling of a Nickel specimen, according to Jeschar et al.[17]

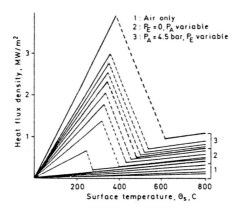

Fig. 16. Heat flux density versus surface temperature at different fluid pressures, according to Archambault et al.[18] P_A = air pressure; P_E = water pressure.

4 GAS QUENCHING IN VACUUM FURNACES

The rapid cooling after austenitisation in vacuum furnaces is achieved by blasting an inert gas around the vacuum furnace and over the workpieces. A variety of vacuum furnace designs have evolved, resulting in a range of cooling rates by varying gas pressure, gas velocities and gas flow patterns. In the cooling of any steel components the process is limited by:

- gas parameters which control the rate of heat removal from the surface of the component, and
- component parameters which control the rate of heat transfer within the component from the centre to the surface.

22 Quenching and Carburising

Gas parameters dominate the cooling rate in small diameter components while component parameters dominate the cooling rate in large diameter components. Both types of parameters must be considered.

Figure 17 (a–f) shows, according to Radcliffe,[19] the influence of the most important parameters for gas quenching in vacuum furnaces. The effect of gas temperature on the cooling of 25 mm diameter steel slugs is shown in Fig. 17a. In the initial cooling period the gas temperature has only a minimal effect on the temperature of the slug. However, after this initial period the component cooling rate becomes increasingly sensitive to changes in gas temperature, the cooling rate decreasing as the gas temperature increases.

The effect of gas velocity is shown in Fig. 17b. Local gas velocities increased around 25 mm diameter component, by increasing the gas flow rate from 2.1 $m^3 s^{-1}$ to 3.5 $m^3 s^{-1}$.

The effect of increased gas pressure on the cooling of 25 mm diameter × 48 mm specimens is shown in Fig. 17c.

The heat transfer coefficient α for a given gas depends on the local gas velocity (V) and gas pressure (P) according to the following formula:

$$\alpha = C\,(VP)^m \tag{8}$$

where m and C are constants that depend on furnace design, component size and load configuration ($m = 0.6$–0.8).

The effect of heat transfer coefficient on the cooling rate of a 25 mm diameter steel slug is shown in Fig. 17a.

Equal increases, in either local gas velocity or pressure, have the same effect on the heat transfer coefficient and hence on the cooling rate of a component. The increase of either gas velocity or pressure affects the design of the blower and the power required to recirculate gases. Doubling of the gas velocity increases the blower power by a factor of eight, while doubling the gas pressure only increases the blower power by a factor of two. Therefore, vacuum furnaces for heat treatment have been built depending on the quenching intensity required: with 6, 10 or even 20 bar overpressure, and the following α values may be obtained:

1 bar	(N_2 circulation)	$\alpha = 100$–150	$W\,m^{-2}\,K$
6 bar	(N_2 rapid cooling)	$\alpha = 300$–400	$W\,m^{-2}\,K$
20 bar	(He/N_2 rapid cooling)	$\alpha = cca\ 1000$	$W\,m^{-2}\,K$

High α values are necessary for through hardening of thick cold work or hot work tool steel components; for gas quenching of medium alloyed steels and also to guarantee a homogeneous quench throughout a compactly loaded batch of parts.

The heat transfer coefficient is also a function of gas properties, as can be seen from Fig. 17d, which shows the effect of four of these gases on the cooling of 25 mm diameter steel slugs. The physical properties of these gases are listed in Table 2.

Density, dynamic viscosity, specific heat capacity and heat conductivity are important properties for the heat transfer coefficient of a gas or a gas mixture. As Fig. 17d shows, argon gives poor cooling rates and is therefore not used. Hydrogen is also not used because of high explosion risk. Nitrogen has hitherto been mostly

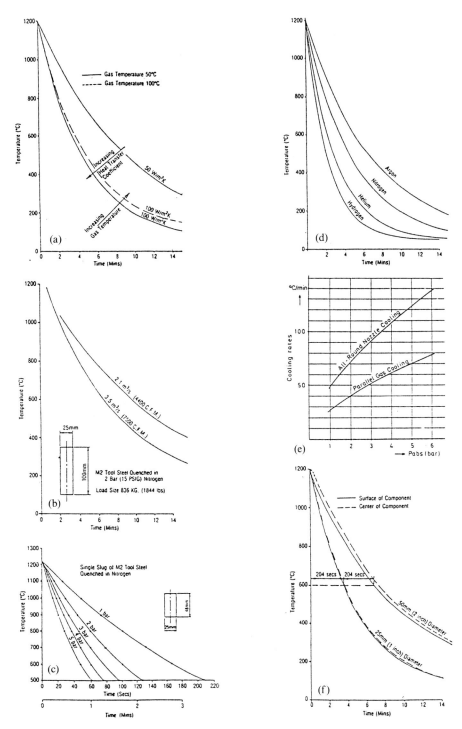

Fig. 17. Influence of the following parameters on cooling curves or cooling rate for gas quenching in vacuum furnaces: (a) gas temperature, and heat transfer coefficient; (b) gas velocity; (c) gas pressure; (d) gas properties; (e) gas flow pattern (furnace design); (f) size (mass) of work-piece.

Table 2. Physical properties of gases

	Chemical symbol	Density at 15°C and 1 bar kg/m³ ρ	Specific heat capacity kJ/kgK C_p	Heat conductivity λ W/mK	Dynamic viscosity η Ns/m²
Argon	Ar	1.6687	0.5204	177×10^{-4}	22.6×10^{-6}
Nitrogen	N_2	1.170	1.041	259×10^{-4}	17.74×10^{-6}
Helium	He	0.167	5.1931	1500×10^{-4}	19.68×10^{-6}
Hydrogen	H_2	0.0841	14.3	1869×10^{-4}	8.92×10^{-6}

used as the cooling gas in vacuum furnaces, because it is much cheaper than helium.

Recently, helium is also used because of its much lower density, higher specific heat capacity and much higher heat conductivity and because of substantial lower blower power required. The optimum heat transfer coefficients have been obtained using a gas mixture of 70–80 vol % He and 20–30 vol % N_2. Because the price of helium is about 10 times higher than that of N_2, a recycling facility is used.

Figure 17e shows how the cooling rate depends on gas flow pattern, i.e. on the design of the vacuum furnace. All-round nozzle cooling system gives higher cooling rates and more uniform cooling than a parallel gas cooling system.

Figure 17f shows the effect of bar diameter on the cooling curves. At the surface of the component the cooling rate is inversely proportional to the component diameter. By increasing the diameter by a factor of two, the cooling rate is decreased by a factor of two. The cooling rate required to achieve hardened steel structure in a particular component depends on one hand on the cross-section size, and on the other on the steel type used. Therefore, continuous cooling transformation (CCT) diagrams are of great help.

5 NEW DEVELOPMENTS IN HOT SALT-BATH QUENCHING

Martempering (marquenching) in a hot salt bath, whose temperature is usually slightly above the M_s temperature of the steel concerned, is a well known quenching process that is successfully applied in order to minimise internal stresses and distortion. In this quenching process there are three main parameters which determine the quenching intensity: the temperature of the bath itself, the agitation rate of the salt and addition of small quantities of water.

In order to apply this process also to steel grades of lower hardenability (non-alloyed grades) or to workpieces of bigger cross-section, the agitation rate of the salt should be increased and/or up to 2 vol % of water should be added to the salt bath at its temperatures between 180°C and 250°C (less water at higher

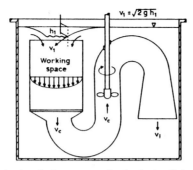

Fig. 18. Scheme of the salt circulation system in the hot salt bath for martempering.

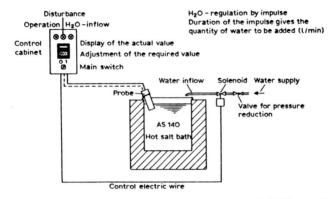

Fig. 19. Scheme of the electronic device (Proprietary: Degussa A.G. Hanau, Germany) for automatic continuous addition of some percentage of water into the hot salt.

temperatures). A very effective salt circulation system with violent downward flow of liquid salt within the working space is shown in Fig. 18. By using a two-speed electro-motor driven propeller pump, agitation rates of 0.3 and 0.6 m/s in the working space are possible.

By adding water into a salt bath whose temperature is higher than the boiling temperature of water, the main problem is how to keep this added amount of water constant, because of evaporation.

Figure 19 shows a newly developed device for automatic continuous addition of water into the hot salt bath.

Provided the temperature of the salt bath and its agitation rate are constant, its quenching intensity can be maintained within close tolerance by the application of the mentioned device. The principle of its function is based on the simulation of cooling of a workpiece. The device consists of three main parts: the probe for temperature measurement, the electronic control and the system for adding water.

26 *Quenching and Carburising*

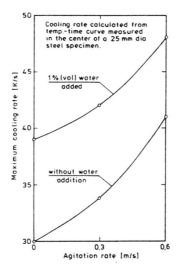

Fig. 20. Maximum cooling rate of a hot salt bath (Degussa AS-140) of 200°C as a function of the agitation rate and percentage of water added.

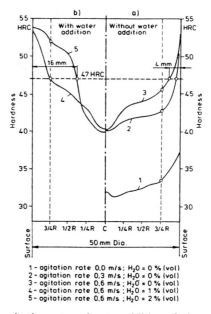

1 - agitation rate 0.0 m/s ; H_2O = 0 % (vol)
2 - agitation rate 0.3 m/s ; H_2O = 0 % (vol)
3 - agitation rate 0.6 m/s ; H_2O = 0 % (vol)
4 - agitation rate 0.6 m/s ; H_2O = 1 % (vol)
5 - agitation rate 0.6 m/s ; H_2O = 2 % (vol)

Fig. 21. Influence of the agitation rate and water addition of a hot salt bath (Degussa-AS-140) of 200°C on hardness distribution after quenching 50 mm diameter × 200 mm bars made of AISI-4140 (batch No 89960).

The probe which is permanently immersed into the salt bath is heated every 3 minutes, once by induction to 500°C. After reaching this temperature, the heating cycle is interrupted and the probe cools down to the salt bath temperature. The period of time needed to cool the probe, which depends on the quenching intensity of the salt bath, is measured and compared to a preset value. If this period of cooling becomes longer, indicating that the quenching intensity of the salt bath has decreased, a signal will be given to the solenoid valve to add more water.

In order to evaluate the influence of water addition and the agitation rate on the quenching intensity of the hot salt bath (Degussa AS-140) at 200°C, temperature vs. time curves have been measured in the centre of a 25 mm diameter steel specimen, and relevant cooling rate curves have been calculated.

Figure 20 shows the influence of the agitation rate as well as of 1% (vol) water addition on maximum cooling rates achieved. Figure 21 shows the influence of the agitation rate of the hot salt bath: (a) without water addition, and (b) with water addition, on hardness distribution after quenching 50 mm diameter × 200 mm bars made of AISI-4140 steel. As can be seen from Fig. 21(a), the greatest increase in hardness has been achieved between a motionless and an agitated bath, but further increase in hardness and especially in the achieved depth of hardening is achieved by adding up to 2% (vol) of water (see Fig. 21(b)). By taking the hardness of 3/4 R as the representative value it is evident that, with good agitaiton and 2% (vol) water addition, an increase of about 19 HRC is achieved in comparison with the same hot salt bath without agitation and without water addition. The depth of hardening at a level of 47 HRC is four times larger with 2% (vol) water addition than without water addition if all other parameters are equal.

6 INTENSIVE QUENCHING

Intensive quenching is not a new invention. It has been used for a long time in the automotive and aircraft industry in the USA,[21] as well as in the automotive industry in the Soviet Union.[22] However, its benefits are not widely known and only in recent years investigations and calculations have been carried out to explain its theoretical background.

According to Kobasko[23], in recent years intensive quenching has been more and more applied because it was found that with forced heat extraction, additional material strengthening with simultaneous improvement of ductility, without quenching cracks, has been observed.

It is, hitherto, a general belief that the 'critical cooling rate' (represented by the cooling curve just passing by the nose of pearlitic or bainitic transformation start – see Fig. 22) which is necessary to obtain 100% martensite, should not be overstepped, because of the risk of cracking. Therefore, only few investigations have dealt with quenching processes involving cooling rates higher than critical. In one of them[24] it was found that the probability of quenching cracks occurrence, in relation to increasing cooling rate (in the temperature region of martensite formation), first becomes higher and after reaching a maximum decreases to zero

28 Quenching and Carburising

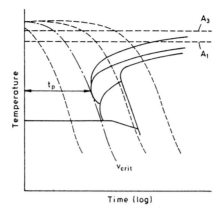

Fig. 22. Cooling curve corresponding to the 'critical cooling rate' (v_{crit}). t_p = incubation period of pearlitic transformation.

(see Fig. 23). The strength of quenched specimens having martensitic microstructure shows the opposite course. By increasing the cooling rate, the strength values first decrease to a minimum and by further increase of the cooling rate they again become higher. These results show that, at cooling rates substantially higher than critical in the region of martensite formation, strengthening of martensite occurs. According to Ivanova[25], the reason for this is the emerging of a huge amount of dislocations.

The explanation of the very astonishing fact that quenching cracks do not occur at very high cooling rates can be found in Kobasko and Morganjuk[26]. The process of quenching has been simulated with the help of the mathematical model which included the equation of non-stationary heat conduction and equations of the theory of elastic-plastic flows with kinetic strengthening under the relevant boundary conditions. It has been found that, by increasing cooling rates, the

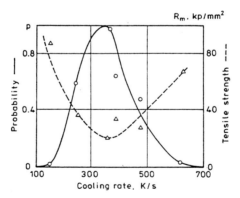

Fig. 23. Probability of quenching cracks occurrence and strength values of quenched specimens of AISI-5140 vs. cooling rate in the temperature region of martensite formation.[24]

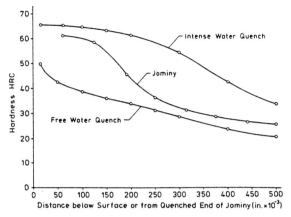

Fig. 24. Typical advantage of intense quenching with water jets for 75 mm diameter bars made of AISI-1045 steel. $H > 5.0$.[21]

tensile residual stresses (tangential and axial) first become greater and then at higher cooling rates decrease and change to compressive residual stresses. The greater the cooling rate, the higher the compressive stresses at the surface of the part being quenched.

With intensive quenching high temperature gradients at the surface are obtained, which determine the depth of martensite layer. An optimum depth of martensite layer in each case corresponds to the maximum compressive stresses. These compressive stresses in the surface layer are, of course, balanced by the tensile stresses in the core. Results of mathematical modelling show that the maximum of compressive stresses in the surface layer corresponds to the maximum of tensile stresses in the core. Therefore, the intensive quenching process should be interrupted at the moment of reaching the maximum tensile stresses in the core (which, according to some investigations, is at 450–500°C core temperature). Based on these results, a new quenching technology for alloyed steel grades has been proposed. It consists of:

- very intensive heat extraction until the moment when maximum tensile stresses in the core and maximum compressive stresses in the surface layer are reached, and then:
- isothermal holding at about martensite start temperature ($M_s \pm 20°C$) until full transformation of austenite in the core.

The advantage of this technology is the avoidance of quenching cracks formation in conditions of highly intensive heat extraction, when quenching alloyed steels. A high and uniform hardness is achieved in the surface layer in which high compressive stresses are created. A combined process, which is used for high carbon steel grades, is also possible. It consists of:

– very intensive heat extraction in the first stage until the temperature (below M_s) is reached at which about 25–30% of martensite will be formed;

30 *Quenching and Carburising*

– holding the parts in the air at this temperature until temperature equalisation in the cross-section;
– intensive cooling in the second stage, during transformation of the rest of martensite, in order to increase the strength, but simultaneously also plastic properties of the material.

Intensive quenching certainly calls for adequate facilities, and the following quenching media may be used: pressurised water-jets; water solutions of such additives, the boiling temperature of which can be maintained at the required temperature (applicable in the first above mentioned stage); liquid nitrogen (applicable in the second above mentioned stage). It is worthy to be noted that, according to Kobasko[24], the cooling rate at 150°C is five times greater when immersing the part in liquid nitrogen than by cooling it with a water jet of 10 bars pressure!

As a result of the development of high residual compressive stresses in the surface layer of intensively quenched parts, excellent fatigue resistance is obtained. Higher hardness and bigger depth of hardening, as is shown in Fig. 24, can also result from intensive quenching. This offers the possibility in some cases, to use a cheaper carbon steel instead of an alloyed one.

According to Kern[21], the reasons for using intensive quenching may be one or a combination of the following:

- To achieve microstructures, in the surface of a part and in some instances below the surface, that are essentially 100% martensitic.
- To develop high temperature gradients and to obtain high residual compressive stresses.
- To accomplish both above mentioned uniformly, on a part in prescribed areas.
- To obtain the maximum hardened depth for a given amount of alloy, to minimize steel cost.
- To insure the maximum hardening in critical areas of parts such as in the root fillets of gear teeth.
- To minimise heat treating distortion and maximise its uniformity.

7 PROSPECTS

In order to promote the present practice in quenching more closely to the field of 'intelligent material processing', the following trends should be followed:

(a) Better understanding of stress analysis; more investigations and applications of results in this field as a vital prerequisite for distortion control.
(b) To investigate and apply new possibilities of higher heat transfer rates (intensive quenching methods), and develop new quenching technologies and quenching facilities, both for single part quenching and for batch quenching.
(c) To realise the automatic control of quenching intensity (with adequate sensors) during the whole quenching process, for different quenching techniques.

(d) Prediction and modelling of metallurgical transformations, hardness (strength) distribution and the stress-strain state after quenching, is an imminent task. Therefore, adequate mathematical and computer-aided research work is necessary to clarify many effects and correlations still unknown or uncertain in this very complicated field of quenching.

REFERENCES

1. 'HIP benefits augmented via "HIP quenching"', *Advanced Materials & Processes, 3/91*, **139**, 37–38.
2. D. Persampieri, A. SanRoman and P.D. Hilton: 'Process Modelling for Improved Heat Treating', *Advanced Materials & Processes, 3/91*, **139**, 19–23.
3. P. Archambault, G. Didier, F. Moreaux, and G. Beck: 'Computer Controlled Spray Quenching', *Metal Progress, October 1984*, p. 67–72.
4. S. Segerberg: 'Technische Eigenschaften und gesundheitliche Verträglichkeit von Polymerabschreckmitteln', *HTM 42 1987*, **1**, 50–54.
5. K.E. Thelning: Proceedings of 5th IFHT Congress on Heat Treatment of Materials, Budapest 1986, vol. 3, pp. 1737–1759.
6. G.E. Totten, M.E. Dakins and R.W. Heins: 'Cooling Curve Analysis of Synthetic Quenchants – A Historical Perspective', *J. Heat Treat. 1988*, **6**, 87–95.
7. M.E. Dakins, G.E. Totten and R.W. Heins: 'Cooling Curve Shape Analysis Can Help Evaluate Quenchants', *Heat Treating*, December 1988, 38–39.
8. G.E. Totten, M.E. Dakins, K.P. Anathapadmanabhan and R.W. Heins: 'Cooling Rate Curve Area: A New Measure of Quenchant Performance', *Heat Treating*, December 1987, 18–20.
9. C.E. Bates and G.E. Totten: 'Quantifying Quench – Oil Cooling Characteristics', *Advanced Materials & Processes 3/91*, 25–28.
10. B. Liščić: 'Der Temperaturgradient auf der Oberfläche als Kenngrösse für die reale Abschreckintensität beim Härten', *HTM, 4/1978*, **33**, 179–191.
11. B. Liščić: 'A New Method to Measure and Record the Quenching Intensity in Workshop Practice', Proceedings of the 6th International Congress on Heat Treatment of Materials, Chicago 1988. Published in *Heat Treatment and Surface Engineering*, New Technology and Practical Applications, ASM 1988, 157–166.
12. N.I. Kobasko: 'Thermal Processes in Steel Quenching', *Metallovedenie i termičeskaja obrabotka metallov, 1968*, No 3.
13. C.E. Bates: 'Predicting Properties and Minimizing Residual Stress in Quenched Steel Parts', *J. Heat Treating, 1987*, **6** (1), 27–45.
14. M. Živković: 'Kriteriji intenziteta ohladivanja i vlastite napetosti pri kaljenju', Diplomski rad – Fakultet strojarstva i brodogradnje – Sveučilišta u Zagrebu, 1991.
15. H.M. Tensi, Th. Künzel and P. Stitzelberger: 'Benetzungskinetik als wichtige Kenngrösse für die Härtung beim Tauchkühlen', *HTM (1987)* **42** (3), 125–131.
16. N. Shimizu and I. Tamura: 'An Examination of the Relation between Quench-hardening Behaviour of Steel and Cooling Curve in Oil', *Transactions ISIJ, 1978*. **18**, 445–450.
17. R. Jeschar, R. Maass and C. Köhler: 'Wärmeübertragung beim Kühlen heisser Metalle mit verdampfenden Flüssigkeiten unter besonderer Berück-sichtigung des Spritzstrahlens', Proceedings of the AWT-Tagung Induktives Randschichthärten, 23–25, March 1988, Darmstadt, pp 69–81.
18. P. Archambault, G. Didier, F. Moreaux and G. Beck: 'Computer Controlled Spray Quenching', *Metal Progress, October 1984*, pp. 67–72.

19. E.J. Radcliffe: 'Gas Quenching in Vacuum Furnaces – A Review of Fundamentals', *Industrial Heating, November 1987*, pp. 34–39.
20. G. Wahl: 'Entwicklung und Anwendung von Salzbädern für die Wärmebehandlung von Einsatzstählen', *Durferrit-Technische Mitteilungen*, Degussa A.G.
21. Roy F. Kern: 'Intense Quenching', *Heat Treating, September 1986*, pp. 19–23.
22. N.I. Kobasko: 'Zakalka stali v židkih sredah pod davleniem', *Naukova Dumka*, Kiev 1980.
23. N.I. Kobasko: 'Povišenie dolgovečnosti i nadežnosti raboti izdelii pri ispoljzovanii novih sposobov zahalki stali', *Metallovedenie i termičeskaja obrabotka metallov (1989)*, **9** 7–14.
24. N.I. Kobasko: 'O. putjah upročnenia stali na osnove intensifikaciji procesov teploobmena v oblasti martensitnih prevraščenii', *Metalli*, No. 1, 146–153, Izvestija akademii nauk SSSR, Moskva 1979.
25. V.S. Ivanova: 'Rolj dislokacii v upročnenii i razrušenii metallov', *M. 'Nauka'*, 1974.
26. N.I. Kobasko and V.S. Morganjuk: 'Issledovanie teplovogo i napraženno-deformirovannogo sostojanija pri termičeskoj obrabotke izdeli energomašinostroenija', Kiev: Znanie, 1983, 16.

2

Measurement and Evaluation of the Quenching Power of Quenching Media for Hardening

J. BODIN AND S. SEGERBERG

IVF – The Swedish Institute of Production Engineering Research, Mölndalsvägen 85, S-412 85 Gothenburg, Sweden

ABSTRACT

Distinction is first made between the concepts **cooling power** and **hardening power**. Various methods of testing the cooling power are presented together with existing and proposed standards for testing of quenchants.

A number of methods have been proposed for testing the hardening power, which take account of the material (steel) to be hardened. Some of these methods are mentioned.

The interrelation between the quenchant and steel being hardened has been described in a number of ways over the years. Recently, new ways of determining the hardening power, based on information from the cooling curve of the quenchant, as recorded in a standardised test, have been proposed.

With knowledge of the hardening power of available quenchants, selection of the most suitable one for each application is simplified. A list of several commercial quenchants with their hardening power in relation to carbon and low-alloyed steels is presented.

1 INTRODUCTION

Quenching is the most critical part of the hardening process. The quenching process has to be designed so as to extract heat from the hot workpiece at such a rate as to produce the microstructure, hardness and residual stresses desired.

Basic research on the quenching process has been carried out since the beginning of this century. During the last decade, important progress has been made in developing deeper understanding of the quenching process and its influence on the properties of workpieces being hardened.

The rapid evolution of computer hardware both offers tools to facilitate testing and encourages development of software to predict the results of hardening. This development has stressed the importance of better understanding of the interrelation between the quenchant and the material being hardened as well as of standardised methods for testing quenching media.

34 Quenching and Carburising

Before entering into details about methods of testing and evaluation of quenchants, it is essential to make a distinction between the two concepts: **cooling power** and **hardening power** of quenchants. According to *Metals Handbook*,[1] the cooling power is the 'thermal response' of the quenchant or the rate of heat removal from a specimen, usually instrumented (see Chapter 2), while the hardening power is the 'metallurgical response' of steel specimens or the ability of the quenchant to develop a specified hardness in a given material/section size combination (see Chapters 4 and 5).

2 METHODS OF TESTING THE COOLING POWER OF QUENCHING MEDIA AND QUENCHING SYSTEMS

Grossmann et al.[2] introduced the ***H* value** (the 'heat transfer equivalent') for the cooling power (also called the 'severity of quench') of quenching media. The H value is defined as:

$$H = h/2k$$

where

h = the mean value of the heat transfer coefficient throughout the entire quench process

k = the thermal conductivity of the material

According to Grossmann, the H value for stationary water at room temperature is about 1.0 (cm^{-1}), while, consequently, slower-cooling quenchants have H values less than 1. Sometimes the H value is determined in comparison tests, where the relative cooling time between two temperatures, e.g. 700°C and 550°C or the cooling rate at a relevant temperature, e.g. 700°C, is measured for stationary water and the quenchant of interest. Thus, the H value takes little or no consideration of the fact that both the the heat transfer coefficient and the thermal conductivity vary with temperature. Therefore, a quenchant with a higher H value will not always produce a higher hardness when real steel workpieces are quenched. To be able to evaluate the result of hardening from the cooling power, it is obvious that the complete cooling process has to be considered, as in cooling curve (or thermal gradient) testing. This is dealt with in Chapter 4.

2.1 COOLING CURVE TESTING

This method is the most useful for testing the cooling power of quenching media and quenching systems, since the complete cooling process is recorded. The test is performed by quenching a test-piece (probe) with a thermocouple embedded at some point, usually at its geometric centre or at (or near) the surface, and monitoring the cooling process with a temperature-measuring device.

Figure 1 shows a typical cooling curve, where the three phases normally

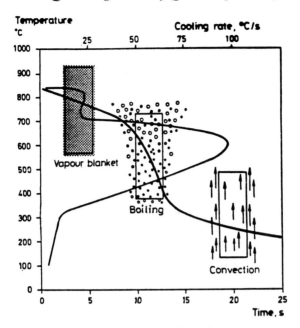

Fig. 1. Typical cooling curve for a quenchant with its three phases.

appearing when quenching in liquid media are clearly identified. In the figure, the cooling rate vs. temperature is also shown, as is common practice.

The probe *material* is usually austenitic stainless steel, nickel or nickel alloys or silver, all of which are free from transformation effects, or of carbon or alloyed steel. The probe *shape* is usually cylindrical. For many years, silver sphere probes were often used but, mainly for manufacturing reasons, cylindrical shapes are preferred today.

The cooling curves recorded are influenced, naturally, by the thermal properties of the probe material, the probe size and the position of the thermocouple in the probe. On the one hand, the sensitivity of measurement increases with increasing thermal conductivity of the probe material and decreasing probe size (when temperature is measured at the probe centre), or if temperature is measured at (or near) the probe surface. On the other hand, manufacture of the probe and design of the temperature monitoring system are more critical when probe sensitivity increases. Therefore, choice of probe material and size as well as type and position of the thermocouple have to be a compromise between sensitivity, manufacturing costs and probe life.

Cooling curve testing is now the subject of international standardisation and various national standards exist (see below). Equipment for testing, both stationary for laboratory use and portable for testing in the workshop, is available commercially.

36 *Quenching and Carburising*

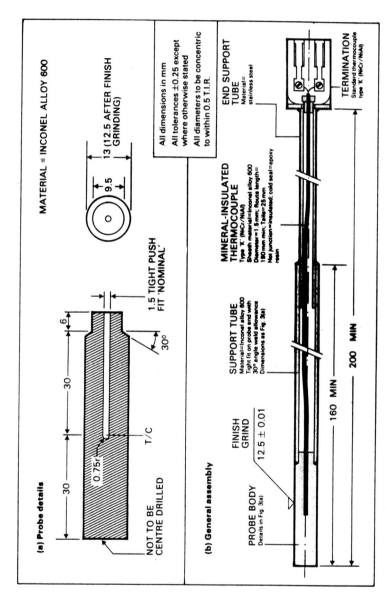

Fig. 2. Test probe according to the proposed international standard (ISO/DIS 9950 draft).[3]

2.1.1 International standards

The International Federation for Heat Treatment and Surface Engineering (IFHT) and its Scientific and Technical Aspects on Quenching Sub-Committee is engaged in preparing proposals for international standards concerning quenchant testing. Different methods have been compared in round-robin tests performed in a number of countries. Based on this, proposals for standards are being submitted to the International Organisation for Standardisation (ISO).

The first proposal involving laboratory testing of *oils*[3] has been submitted to ISO and is now being reviewed by its TC 28 Technical Committee. It involves testing with a *$\phi 12.5 \times 60$ mm cylindrical probe* made of *Inconel 600 (UNS N06600)*. Figure 2 shows the design of the test probe.

Testing is done from 850°C in a 2000 ml sample of stationary oil. Results of testing are expressed by (a) the temperature/time curve and the cooling rate/temperature curves, and (b) the following data read off from these curves:

- cooling time to 600°, 400° and 200°C;
- maximum cooling rate;
- temperature at which the maximum cooling rate occurs;
- cooling rate at 300°C.

The draft proposal is based on a specification elaborated by Wolfson Heat Treatment Centre, Great Britain.[4] It has already reached widespread use, e.g. in Europe and the USA.

A few countries have standards for testing with silver probes (see below). Therefore, the IFHT Sub-Committee recommended that, as a *second-choice ISO standard* for testing of oils, testing with a *silver probe* according to the French national standard (see below) should be accepted. This will also be considered within the ISO committee.

As far as testing of *water-based polymer quenchants* is concerned, the IFHT Sub-Committee has organised round-robin tests which have been performed recently. The results are now being analysed. A draft proposal is expected shortly. In contrast to oils, testing of polymers has to be done with agitated quenchants in order to receive reproducible results. Therefore, the draft will also include a specification of the agitation system.

2.1.2 Some national standards

France. The French standard for testing of oils, NFT 60178,[5] is based on a $\phi 16 \times 48$ mm silver cylinder probe. Testing is done from 800°C in a 700 ml sample of stationary oil. In addition, there is a standard, NFT 60512, which classifies quenching fluids in product families and the principal fields of application.

P.R. China. The Chinese standard GB 9449-88[6] is identical with the French standard as concerns probe size and material, and probe and oil temperatures. The main difference in comparison with this standard is that a more sensitive probe

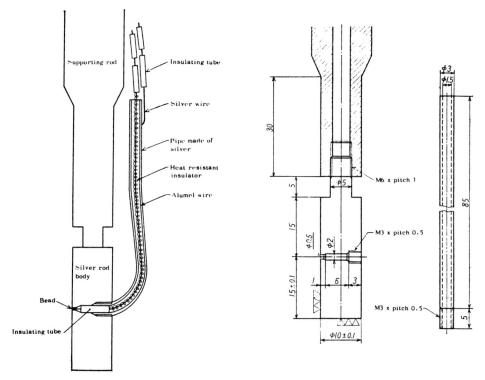

Fig. 3. Test probe according to the Japanese standard JIS K 2242-1980.[7]

thermocouple (0.5 mm instead of 1.0 mm dia.) is specified. The Chinese standard applies to both oil-based and water-based quenchants.

Japan. The Japanese standard JIS K 2242[7] contains specification of heat treating oils and their classification as well as a method of testing their cooling performance. In this, a $\phi 10 \times 30$ mm *silver cylinder* equipped with a thermocouple at its surface is used, see Fig. 3. Testing is done from 810°C in a 250 ml sample of the oil. The result is expressed by the characteristic temperature (i.e. the transition temperature between film boiling and nucleate boiling) and the cooling time from 800° to 400°C derived from the cooling curve.

Recently, the standard has been subject to modification, mainly in order to lengthen probe life. Other probe designs have been considered, including those where the thermocouple is positioned in the centre of the probe. It has been decided, however, to stay with the existing design, essentially in order not to impair sensitivity of measurement.

USA. A draft specification has been elaborated recently by the ASM Quenching & Cooling Committee and is being discussed with the American Society for Testing and Materials (ASTM) for possible standardisation. The specification is similar to the ISO draft. Beside ASTM, some other competent US organisations are discussing standardisation of test methods for quenchants.

Great Britain. Although not formally standardised, the Wolfson specification,[4] which is similar to the ISO draft, has been accepted as a *de facto* standard within the British industry since its publication in 1982. In the first drafts from the mid-1970s, stainless steel was specified for the probe but later a change to Inconel 600 was made in order to increase the resistance to oxidation and corrosion.

Comments on the standards. The Wolfson method (and, thus, the ISO/DIS 9950 draft) can be characterised as an *engineering* method, where probe life, reproducibility and, to some extent, similarity with quenching of steel have been considered the most important. The Japanese method in particular, and also the French, can be considered as *scientific*, where sensitivity has been considered the most important.

2.2 THERMAL GRADIENT TESTING

To be able to calculate the temperature within a workpiece being quenched, the heat transfer coefficient, or the heat flux, at the surface has to be determined. This can be calculated from measurements with thermocouples at or near the surface of the workpiece (i.e. cooling curves). However, as the heat flux from the workpiece is proportional to the *thermal gradient* at the surface, it can be determined directly by measuring the cooling curve at two points, at and near the surface. A method based on this concept has been developed by Liscic.[8] A $\phi 50 \times 200$ mm stainless steel probe is used, having three thermocouples (at the surface, 1.5 mm below the surface and at the centre). The quenching intensity is represented by the heat flux density at the surface as calculated from the thermal gradient. The method is intended to be used also in workshop practice.

2.3 OTHER METHODS FOR TESTING THE COOLING POWER

The following methods have also been used to determine the cooling power of quenchants. These methods produce only small fractions of the information contained in cooling curves, as with the H value, and so their applicability is limited. Their advantage is the simplicity. However, one has to be very careful when interpreting results from measurements with these methods. For further details see for example Refs. 1 and 9.

Houghton quench test (system Meinhardt). This method[10] is a simplified version of the cooling curve test. A test-piece of carbon steel, normally with thermocouple wire, which together with the steel probe and support tube forms the thermocouple, is immersed in the quenchant. The time to cool within a certain temperature interval is determined with a built-in stop watch. Thus, two points on the cooling

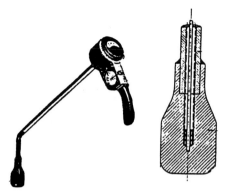

Fig. 4. Houghton quench test apparatus (system Meinhardt).[10]

curve can be determined in each test. The result is expressed as the time to cool between two fixed temperatures, normally from 700° to 400°C (or 300°C). (The probe thermocouple can, of course, also be connected to a temperature recording equipment so as to monitor the complete cooling curve, even if this is not the intended use). Figure 4 shows the equipment.

The diameter of the probe is normally 55 mm, but can be chosen between 20 mm and 80 mm to correspond to different dimension of workpieces to be hardened. In addition to carbon steel, austenitic stainless steel can also be used as the probe material. In this case, a standard thermocouple must be used instead of the arrangement in Fig. 4.

The magnetic test. This makes use of the fact that magnetic metals lose their magnetic properties above the Curie temperature. In this test a nickel sphere is heated to 835°C and quenched in a sample of the quenchant. The time required to cool to the Curie temperature, which is 355°C, is recorded. This time is a measure of the cooling power of the quenchant. By choosing other alloys with different Curie points, additional points on the cooling curve can be determined. The method, also known as the GM Quenchometer test, is standardised in the USA (ASTM 3520).

Hot wire test. In this test, a resistance wire is heated by means of an electric current in a sample of the quenchant. The current is increased steadily and the cooling power is indicated by the maximum steady current reading, as measured by an ammeter.

The main disadvantages are that comparison of quenchants is only made in the higher temperature range and that no information about the vapour phase characteristics and the transition temperature to boiling can be obtained (a stable vapour phase can not be achieved with a wire).

Interval test. In this test, also known as the 5-s test, a heated bar of metal (usually stainless steel) is immersed for 5 s in a sample of the quenchant which is contained in an insulated container. The increase in temperature is noted. The process is repeated for a series of bars. Finally, a bar of identical size and material is fully quenched in a new sample of quenchant of the same volume. The cooling power is expressed as the rate between the average rise in temperature for the 5-s quench bars and the rise for the fully-quenched bar.

The method is simple and requires no special equipment. It can be used for determination of gross changes in cooling power. However, as with the interval test, measurement is only made in the high temperature range.

A similar method has been developed from the interval test, where consideration is taken of the cooling power at lower temperatures. Here, the rise in temperature is determined after various immersion times, or continuously.

3 METHODS OF TESTING THE HARDENING POWER OF QUENCHING MEDIA

3.1 IMMERSION QUENCH TESTS

These tests involve immersion of heated steel specimens in samples of the quenchant or in the quench tank itself. Some common practices and standards are presented below.

Cylindrical specimens. Testing with cylindrical specimens is common practice in most heat treatment shops. The dimensions and material of the specimens should be chosen with regard to the products to be hardened and the quenchant to be used. Variation in the hardening power within a quench tank can be monitored by placing specimens at various positions in a batch to be hardened.

Stepped, cylindrical specimens. Instead of using specimens with different diameters, stepped specimen with two (or more) diameters are sometimes used. The French automotive companies Renault and Peugeot, for example, used stepped specimen with two diameters made of 38C2 grade steel for a period of years.[11] It is also obviously used in several other companies.

Wedge-shaped specimen. A method for testing with a wedge-shaped specimen has been developed by the French Association for Heat Treatment (ATTT). The method was submitted for standardisation both in France (NFT 60179 draft)[11] and internationally[12] in 1988.

The specimen is made of 38C2 grade steel (according to the French standard NF A 35-552). The shape of the specimen is shown in Fig. 5. After quenching, the

42 Quenching and Carburising

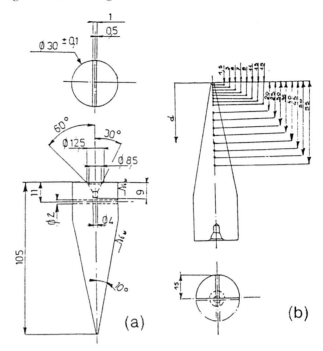

Fig. 5. (a) Wedge-shaped specimen proposed by ATTT, France. (b) Points where hardness is tested.[11,12]

probe is cut into two halves, longitudinally, and the hardness is measured at prescribed points.

3.2 MODIFIED JOMINY END-QUENCH TEST

The Jominy test (ISO 642 and ASTM A 255) is normally used for determination of the hardenability of steels. In this test, the quenching medium is water. According to Ref. 9, the same equipment can be used for evaluation of quenchants as well. In this case, specimen of known hardenability is used and the variables of the quenchant are changed. The quenchant has to be contained in a closed system in order to permit close control. The effect of the variable, whether composition, temperature, concentration or some other factor, is evaluated by the hardness pattern developed on the end-quench specimen.

The advantage would be that testing can be made in equipment that may already be on-site in the company. The main disadvantage would be that the quenching conditions are different from normal hardening in a quench tank (no stable vapour phase will be formed at the probe surface at high temperatures and convection will be greater). However, it could be a relevant method for testing of quenchants for induction hardening or for injection or spray quenching, where the vapour phase does not normally appear.

4 METHODS TO INTERRELATE QUENCHING PERFORMANCE WITH THE PROPERTIES OF STEEL WORKPIECES BEING HARDENED

The interrelation between steel workpieces being hardened and the quenchant has been the subject of many investigations since the beginning of this century and more intensively since the 1930s. Names like Bühler, Grossmann, Bain, Wever, Rose, Peter, Tamura, Kobasko, Hougardy, Atkins, Beck, Thelning and their co-workers are only examples of people who have made important contributions to the understanding of the quench hardening process.

In this chapter, some examples from more recent engineering research are presented, where attempts were made to express the 'hardening power' of the quenchant in a way relevant to the metallurgical response of steel workpieces being hardened. Before going into details about these, some comments are made concerning the possibilities and limitations of using CCT diagrams for evaluation of the quenching process.

4.1 COMMENTS ON THE INTERPRETATION OF COOLING CURVES IN RELATION TO CCT DIAGRAMS

Continuous cooling transformation diagrams (CCT diagrams) are often used to evaluate the metallurgical response of different steels to cooling. Most CCT diagrams are constructed from tests with natural or linear cooling. In industrial quenching in liquids, the cooling rate varies with time and surface temperature of the workpieces. It has been shown by Shimizu and Tamura,[13,14] Thelning,[15] Loria[16] et al that a change in cooling rate during quenching, as in Fig. 1, influences the transformation behaviour of the steel significantly. Shimizu and Tamura[13] also propose a graphical method to find the real transformation behaviour and the critical cooling rate in the CCT diagram, when the cooling rate is changed during continuous cooling.

Thelning[15] has constructed CCT diagrams representing cooling conditions prevailing at the surface and centre of steel bars quenched in water and oil. The diagrams show that the time to transformation is different for water and oil quenching and for the surface and the centre.

Allowance must also be made for the fact that the transformation behaviour depends on such factors as the austenising temperature before quenching, time at elevated temperature (for homogenising) and grain size of the steel.

Consequently, incautious superimposure of cooling curves on CCT diagrams may lead to considerable errors in prediction of the result of hardening.

4.2 SOME METHODS PROPOSED TO CORRELATE COOLING CURVE INFORMATION WITH THE RESULT OF HARDENING

As mentioned in the introduction, computer models of the hardening process are now being developed at several places. A few of them use complete cooling curves or heat transfer data at the surface and work incrementally (in small time steps) to arrive at the final condition successively. Most of the models, however, seem to employ quite simple ways of describing the quenching part of the process and are therefore limited in their application. In this chapter, some simpler methods of correlating the cooling curves with the result of hardening (mostly the hardness) are described.

4.2.1 Cooling rate at a certain temperature or temperature interval

A simple approach in trying to correlate hardening response with the cooling process is to consider the cooling rate at some selected temperature or cooling time in some temperature range. There is general agreement between researchers that a high cooling rate in the temperature region, where the time to transformation to ferrite, perlite and bainite is shortest, is of decisive importance for the hardness and microstructure after hardening.

Table 1 gives examples of temperatures and intervals, where the cooling rate has been correlated to the hardening response, since the turn of the century. Obviously, the temperatures chosen has in some cases, at least, been directly related to the steel being considered.

Atkins and Andrews[17] and Murry[18] have constructed CCT diagrams with the abscissa graded in cooling rate at 800°, 750° or 700°C, depending on the steel composition, and cooling time between 700° and 300°C, respectively.

Figure 6 shows how hardness is correlated with the cooling time between 640° and 400°C for a low-alloy steel. Figure 7 shows similarly how hardness is correlated to the cooling rate at 550°C for carbon steel test-pieces. These figures show good correlation. However, such simple models can only be used with success within certain limits as concerns the cooling characteristics and test-piece dimensions.

4.2.2 The Tamura V value[29]

In this approach, the following steps are identified:

(1) Steels are classified roughly into four types according to the shape of their CCT diagrams (i.e. those for which non-martensitic structures are likely to be ferrite + perlite + bainite, perlite, perlite + bainite or bainite). For each type, a temperature range, X, can be defined, where rapid cooling is required to ensure martensitic transformation.
(2) For a given quenchant, the temperature of the start (T_c) and end (T_d) of boiling is determined by testing with the standardised silver probe (see Fig. 3). These two temperatures are converted into those for steels from graphical data.

Table 1. Temperature ranges and temperatures considered for studies of correlation between cooling time or rate and hardening response

Researchers	Temp. range (°C)	Cooling rate at (°C)
Lechatelier[19]	700–>100	
Benedicks[20]	700–>100	
Mathews, Stagg[21]	650–>370	
Portevin, Garvin[22]	700–>200	
Grossmann et al.[2]	700–>300	
Murry[18]	700–>300	
Deliry et al.[23]	640–>400	
Wever, Rose[24]	800–>500	
Kulmburg et al.[25]	800–>500	
Rogen, Sidan[26]	A_{C3}–>M_S	
Ives et al.[27]		700 and 200
Atkins, Andrews[17]		800, 750 or 700
Segerberg[28]		550

(3) A factor V is defined which is the ratio of the degree of overlap of the two ranges X and T_c–T_d, to X, see Fig. 8.

(4) For a certain section size, the hardening response of a given steel in various quenchants can be predicted from experimentally-determined curves relating the V value with hardness.

4.2.3 The IVF hardening power, HP[30,31]

Based on measurements with the method proposed as an ISO standard for oils (probe according to Fig. 2), a formula has been derived by regression analysis, where the hardening power for *oils* is expressed as a single value, HP:

$$\text{HP} = k_1 + k_2\, T_{\text{VP}} + k_3\, \text{CR} - k_4\, T_{\text{CP}} \qquad (1)$$

where:

T_{VP} = the transition temperature between the vapour phase and the boiling phase (in °C),

CR = the cooling rate over the temperature range 600°–>500°C (in °C/s),

T_{CP} = the transition temperature between the boiling phase and the convection phase (in °C),

k_1, k_2, k_3, k_4 are constants.

For *unalloyed steels* the formula is as follows:

$$\text{HP} = 91.5 + 1.34\, T_{\text{VP}} + 10.88\, \text{CR} - 3.85\, T_{\text{CP}}$$

Hardness ratings of test-pieces of medium-carbon SS 1672 (SAE 1045) steel, hardened in immersion tests in several commercial oils on the Swedish market,

46 Quenching and Carburising

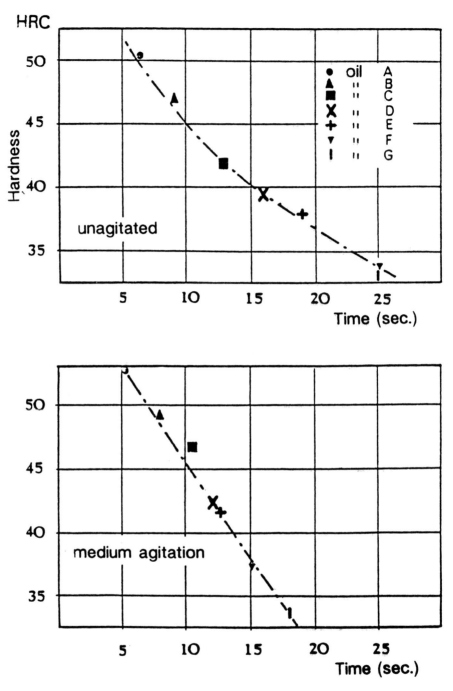

Fig. 6. Correlation between hardness and cooling time from 640° to 400°C in $\phi 16 \times 48$ mm test-pieces of steel 38C2 (0.38% C, 0.5% Cr.)[23]

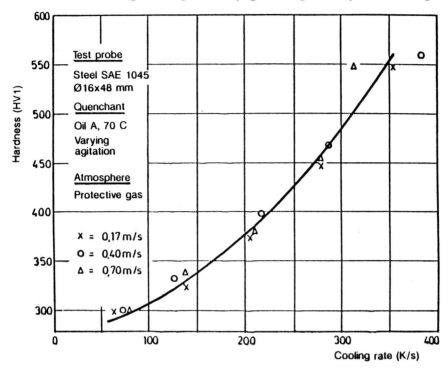

Fig. 7. Correlation between hardness and cooling rate at 550°C for $\phi 16 \times 48$ mm testpieces of steel SS 1672 (SAE 1045).[28]

have been compared with the HP values determined for these oils. The results are shown in Fig. 9, which also shows HP values of some oils on the US market. The definition of the three points on the cooling rate curve are also shown here.

For *alloyed steels*, the coefficients in equation (1) will be different, as will the ranking of oils.

For *polymer quenchants*, a similar approach has been made. However, since the vapour phase is often suppressed, or nonexistent, and the transition between the boiling and convection phases is not as pronounced for polymers, the hardening power has been formulated as follows:

$$HP = k_1 \, CR_F + k_2 \, CR_M - k_3$$

where

CR_F = the cooling rate at the ferrite/perlite nose,
CR_M = the cooling rate at the martensite start temperature,
k_1, k_2 and k_3 are constants.

Figure 10 shows graphically the critical points on the temperature/cooling rate curve for polymers and the relationship between the hardness and the HP values.

48 Quenching and Carburising

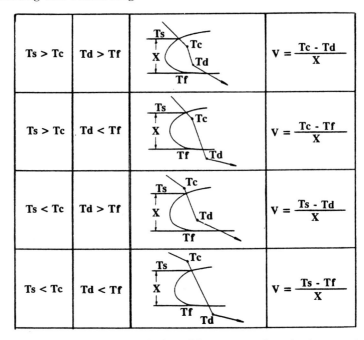

Fig. 8. Calculation of the V value at the four different types of overlap between the boiling phase of the quenchant and the CCT diagram for the steel being hardened.[29]

4.2.4 The Castrol/Renault hardening power, HP, and the Castrol Index, Ci[32]

From extensive tests on oils with probes to the French standard and the ISO draft, as well as with 16 mm diameter probes of Inconel 600 and 38C2 steel with thermocouples at the centre, Deck et al.[32] have developed formulas similar to the IVF formula for oils, but where the hardness achieved in workpieces of 38C2 steel is calculated directly. For the ISO draft probe, the formula is as follows:

$$HP\ (HRC) = 99.6 - 0.17\ \Theta'_2 + 0.19\ V_{400}$$

where:

Θ'_2 = the transition temperature between the boiling and convection phases (in °C),

V_{400} = the cooling rate at 400°C (in °C/s)

The formulas for the other Inconel probe and the steel probe are similar.

In the case of silver, the empricial laws described above could not be applied with sufficient accuracy. Based on thermokinetic reasoning, the following formula was developed, which in fact is independent of the probe material and size:

$$HP\ (HRC) = 3.1\ Ci - 5.7$$

where

$Ci = K'\ V_{max}/(\Theta_{max} - \Theta_t)$

Quenching Power of Quenching Media for Hardening 49

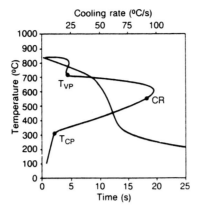

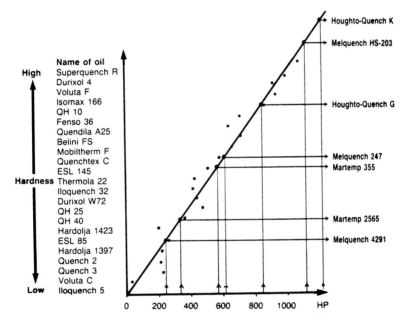

Fig. 9. Predicting the hardening power, HP, of oils: (a) definition of the three points used to calculate the HP value; (b) calculated values of HP matched to a straight line for quenching oils. Performance ranking applies only for hardening of unalloyed steel. HP values determined for some common US oils are also included (right) for comparison (based on measurements made by the respective supplier).[30]

and

$$K' = 11.5 \{(\Theta_{max} - \Theta_t)/V_{max}\}\text{reference}*$$

where

* Reference is made to values for an additive-free base oil, having an H value of 11.5

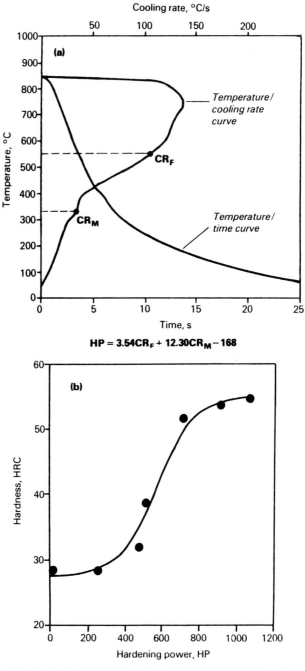

Fig. 10. Predicting the hardening power, HP, of polymer quenchants: (a) definition of the two points used for calculation of the HP value; (b) relationship between the measured hardness on $\phi 16 \times 48$ mm test-pieces of medium-carbon SS 1672 (SAE 1045) steel and the HP value.[31]

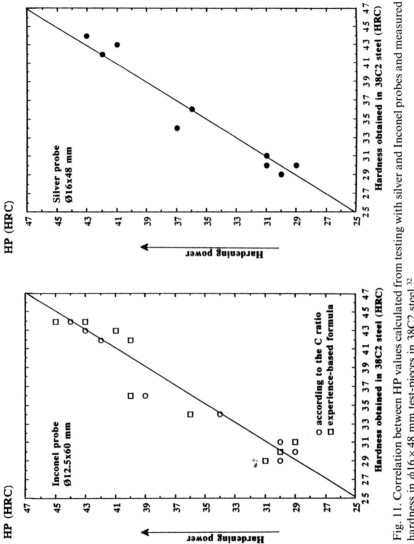

Fig. 11. Correlation between HP values calculated from testing with silver and Inconel probes and measured hardness in $\phi 16 \times 48$ mm test-pieces in 38C2 steel.[32]

52 Quenching and Carburising

V_{max} = the maximum cooling rate (in °C/s),
Θ_{max} = the temperature at which maximum cooling rate occurs (in °C),
Θ_t = the temperature of the quenchant (in °C).

As shown in Fig. 11, the correlation between calculated HP values and hardness measured in steel test-pieces is quite good for both test probes.

4.2.5 Other similar methods

Totten et al.[33] have made an excellent overview of some of the above and several other methods of correlating cooling curves with the results of hardening. These include work by Hilder[34] and Tensi and Steffen[35] on critical transformation times, temperatures and rates, Liscic[36], Thelning[37] and Goryushin et al.[38] on areas under the cooling time and rate curves, Evancho and Staley[39] and Bates[40] on quench factor analysis and Tamura and Tagaya[41] on master cooling curves.

In this context, work by Wünning and Liedtke[42] and Lübben et al.[43] on the QTA method should be mentioned. In this method, the real cooling rate curve is approximated with an artificial one which is more convenient to work with in combination with computer prediction of the result of hardening.

4.2.6 Cooling curves vs. cooling performance in industrial hardening

Cooling curves are normally determined under controlled conditions in the laboratory, although equipment is available on the market today by which the cooling process can be monitored in real quench tanks. The cooling curve, as determined with a standard, or non-standard, test probe, reflects only the cooling characteristics at one point, or one circumference, of the probe. Naturally, the cooling curve characteristics will depend on the size and material of the probe and the position of the thermocouple within the probe. It is also important to note that cooling curves determined with probes made of materials not having phase transformation, especially silver, are different from those of steel.

When quenching real workpieces in a tank, the heat transfer characteristics vary over the surface of the workpiece. The transition between the vapour phase and the boiling phase, for example, takes place at different times at different points on the surface. See, for example, Tensi et al.[44] on wetting kinetics.

The workpieces in a batch also interact with each other, see for example Segerberg and Bodin[45], creating conditions around the workpieces which are difficult, if not impossible, to describe. It has also been found that oxidation of the workpiece surface will change the cooling performance. See, for example, Segerberg[46].

It is therefore of the utmost importance for researchers to work hand-in-hand with practical users of heat treatment in order to be able to transfer results from the laboratory to practice.

5 SUMMARY AND CONCLUSIONS

The ongoing standardisation of methods for testing of quenchants for hardening will bring better tools to the heat treatment industry and to manufacturers of quenchants to monitor and understand the complex quenching process. The possibility of combining immense knowledge from research laboratories, acquired over a century, with practical experience from the workshop floor is better than ever, especially when taking into account the opportunities offered by the revolutionary development in computer technology and applications.

REFERENCES

1. *Metals Handbook*, ninth edn, Vol. 4. Heat Treating, ASM International, USA, 1981, 31.
2. M.A. Grossmann, M. Asimow and S.F. Urban, in *Hardenability of Alloy Steels*. American Society for Metals, 1939, pp. 124–190.
3. *Industrial Quenching Oils – Determination of Cooling Characteristics – Laboratory Test Method*. Draft international standard ISO/DIS 9950, International Organization for Standardization (submitted 1988).
4. *Laboratory Test for Assessing the Cooling Characteristics of Industrial Quenching Media*, Wolfson Heat Treatment Centre, Birmingham, England, 1982.
5. *Drasticité des huiles de trempe. Essai au capteur d'argent*, NFT 60178, Association Francaise de Normalisation, Paris, France.
6. *Test Method for Cooling Properties of Quenching Media*, State Bureau of Standards of the People's Republic of China, Beijing, China.
7. *Japanese Industrial Standard. Heat Treating Oils*, JIS K 2242-1980, Japanese Standards Association, Tokyo, Japan.
8. B. Liscic, in *Proc. 6th Int. Congress on Heat Treatment of Materials. Chicago, USA 28–30 Sept. 1988*, ASM International, USA, 1988, pp. 157–166.
9. *Quenching and Control of Distortion*, H.E. Boyer and P.R. Cary, eds, ASM International, USA, 1988, p. 166.
10. *Houghton Abschreckprüfgerät* (Houghton Quench Test Apparatus). Brochure from Houghton-Hildesheim, Germany.
11. *Fluides de trempe. Determination de la sévérité de trempe d'une installation industrielle*, NFT 60179 (draft specification), Association Technique de Traitement Thermique, Paris, France, 1988.
12. *Quenching Media. Determination of Quenching Severity of an Industrial Facility*, Draft international standard submitted by Association Technique de Traitement Thermique, France, to the International Federation for Heat Treatment and Surface Engineering (IFHT), 1988.
13. N. Shimizu and I. Tamura: *Trans. ISIJ, 1977*, **17**, 469–476.
14. N. Shimizu and I. Tamura: *Trans. ISIJ, 1978*, **18**, 445–450.
15. K.E. Thelning: *Scandinavian J. of Metallurgy, 1978*, **7**, 252–263.
16. E. A. Loria: *Metals Technology, Oct. 1977*, 490–492.
17. M. Atkins and K.W. Andrews, in *BSC Report SP/PTM/6063/-/7/C/*.
18. G. Murry: *Traitement Thermique, 1976*, **108**, 47–54.
19. H. Lechatelier: *Revue de Métallurgie, 1904*, 1.
20. C. Benedicks: *J. Iron Steel Inst., 1908*, 77.
21. J.A. Mathews and H.J. Stagg: *Trans. ASME, 1914*, 36.
22. A.M. Portevin and M. Garvin: *J. Iron Steel Inst., 1919*, 99.
23. J. Delirey, R. El Haik and A. Guimier: *Traitement Thermique, 1980*, **141**, 29–33.
24. F. Wever and A. Rose: *Stahl und Eisen, 1954*, **74**, 749.
25. A. Kulmberg, F. Kornteuer and E. Kaiser, in *Proc. 5th Int. Congress on Heat Treatment of*

Materials, Budapest, Hungary, 20–24 Oct. 1986, Scientific Society of Mechanical Engineers, Budapest, Hungary, 1986, 1730–1736.
26. G. Rogen and H. Sidan: *Berg- und Hüttenmänn. Monatsh., 1972*, **117,** 250–258.
27. M.T. Ives, A.G. Meszaros and R.W. Foreman, in *Heat Treatment of Metals, 1988*, **15,** 11.
28. S. Segerberg: *IVF-skrift 88804*, IVF – The Swedish Institute of Production Engineering Research, 1988.
29. I. Tamura, N. Shimizu and T. Okada: *J. Heat Treating, 1984*, **3,** 335–343.
30. S. Segerberg; *Heat Treating, Dec. 1988*, 30–33.
31. S. Segerberg, in *Heat Treatment of Metals, 1990*, **17,** 67.
32. M. Deck, P. Damay and F. Le Strat, in *Proc. ATTT 90 Internationaux de France du Traitement Thermique, Le Mans, 19–21 Sept. 1990*, Association Technique de Traitement Thermique, Paris, France, 49–70.
33. G.E. Tottten, M.E. Dakins and R.W. Heins: *J. Heat Treating, 1988*, **6,** 87–95.
34. N.A. Hilder: *Heat Treatment of Metals, 1987*, **14,** 31.
35. H.M. Tensi and E. Steffen: *Steel Research, 1985*, **56,** 489.
36. B. Liscic: *Härterei-Tech. Mitt., 1978*, **33,** 179.
37. K.E. Thelning, in *Proc. 5th Int. Congress on Heat Treatment of Materials, Budapest, Hungary, 20–24 Oct. 1986*, Scientific Society of Mechanical Engineers, Budapest, Hungary, 1737–1759.
38. V.V. Goryushin, V.F. Arifmetchikor, A.K. Tsretkov and S.N. Sinetskij: *Met. Sci. and Heat Treatment, 1986*, 709.
39. J.W. Evancho and J.T. Staley: *Metallurgical Trans., 1974*, **5,** 43.
40. C.E. Bates: *J. Heat Treating, 1987*, **5,** 27.
41. I. Tamura and M. Tagaya/ *Trans. Japan Inst. of Metals, 1964*, **5,** 67–75.
42. J. Wünning and D. Liedtke: *Härteri-Tech. Mitt., 1983*, **38,** 149–155.
43. Th. Lübben, H. Bomas, H.P. Hougardy and P. Mayer: *Härterei-Tech. Mitt., 1991*, **46,** 24–34, 155–171.
44. H.M. Tensi, Th. Künzel and P. Stitzelberger: *Härterei-Tech. Mitt., 1987*, **42,** 125–132.
45. S. Segerberg and J. Bodin, *in Proc. 3rd International Seminar: Quenching and Carburising, Melbourne, 2–5 Sept. 1991*, Institute of Metals & Materials Australasia Ltd. Parkville Vic 3052, Australia.
46. S. Segerberg: *IVF-resultat 88605*, Severiges Mekanförbund, Stockholm, 1988.

3
Quench Severity Effects on the Properties of Selected Steel Alloys

CHARLES E. BATES[1]*, GEORGE E. TOTTEN[2]† and
KIMBERLEY B. ORSZAK[2]

[1]University of Alabama at Birmingham, Birmingham, Alabama, USA and [2]Union Carbide Chemicals and Plastics Company Inc., Tarrytown, New York, USA

ABSTRACT

Quenching refers to the rapid cooling of metal parts from the solution-treating temperature, typically from 845 to 870°C (1550 to 1600°F) for steel alloys. Several factors, including the kind of quenchant, the quenchant use conditions, the section thickness of the part and the transformation rates of the alloys being quenched determine whether a part can be successfully quenched and then tempered to produce the desired hardness and strength.

A quench factor, Q, has been derived that interrelates a cooling curve which is dependent on quenching variables, velocity, concentration, temperature effects, etc., part section size and transformation rate data to provide a single number indicating the extent to which a part can be hardened or strengthened.

The physical interpretation of the quench factor and its use to characterise quench media for use in steel heat treating are described. The application of quench factor analysis to different steel alloys and a summary of results obtained in a series of statistical designed experiments conducted to model hardening of steel during oil quenching is also provided.

1 INTRODUCTION

Quenching refers to the rapid cooling of metal from the solution-treating temperature, typically in the range of 845–870°C (1550–1600°F) for steel alloys. Quenching is performed to prevent ferrite or pearlite formation and allow bainite

---

* Charles E. Bates is Research Professor at the Univeristy of Alabama at Birmingham, Birmingham, Alabama, USA.
† George E. Totten is a research scientist and Kimberley B. Orszak is a statistician at Union Carbide Chemicals and Plastics Company Inc. in Tarrytown, New York.

or martensite to be formed. Martensite is the most commonly desired transformation product. After quenching is complete, martensite is tempered to produce the optimum combination of strength, toughness and hardness.

Several factors determine whether a particular part can be successfully hardened, including the heat transfer characteristics of the quenchant, the quenchant use conditions (such as flow velocity and bath temperature), bath loading, section thickness of the part and the transformation characteristics of the specific alloy being quenched. Successful hardening usually means achieving the required hardness, strength or toughness while minimizing residual stress, distortion and the possibility of cracking.

A balance must be obtained between the need to quench quickly enough to retain carbon in the austenite solution and the need to minimise residual stress and distortion in the parts being quenched. Inadequate cooling rates may result in the formation of ferrite or pearlite and a loss of both hardness and strength.

Highly agitated cold water and brine solutions are excellent quenchants in terms of producing martensite. Unfortunately the high cooling rates produce large differences in temperature between thick and thin sections that can cause localised plastic flow, distortion and cracking.

If distortion and cracking are to be minimized, the temperature differences between different areas of a part must be minimised during the quench. This often requires the use of oil or aqueous polymer solutions to mediate heat transfer during the quench. The technical challenges in quench system engineering involve designing systems that will provide uniform heat transfer from the part during quenching and coupling the design with the quenchant medium and operating conditions, e.g. temperature, agitation, etc. to provide the required heat transfer rates throughout the quenching process to provide optimal as-quenched properties.

To adequately analyse a quenchant system, it is necessary to model the heat transfer properties associated wtih both the quenchant and the system. Cooling curve analysis is considered to be the best method of obtaining the information needed. Although cooling curve analysis provides a good method of quantifying heat transfer rates, there is no consensus of opinion on the best method for data interpretation.

Several cooling curve interpretation methods have been recently reported. Some of these methods include the use of:

- rewetting times (Tensi);[1-5]
- an empirical hardening power predictor (Segerberg,[6] Deck and Damay[7]);
- a rigorous analysis of the cooling process (Liscic).[8]

These are good methods with different advantages. In some regards, they can be viewed as complementary.

The objective of this discussion is to describe the use of another quenchant characterisation method – quench factor analysis. This discussion will include a description of the quench factor analysis method, its application to the quenching of different steel alloys and a summary of some statistically designed experiments conducted to model the quenching behaviour of some oil quenchants.

2 QUENCH FACTOR CONCEPTS

2.1 BACKGROUND

The basic hypothesis behind the quench factor concept is that the hardening of steel during cooling can be predicted by segmenting a cooling curve into discrete temperature-time increments and determining the ratio of amount of time the metal was at each temperature divided by the amount of time required to obtain a specified amount of transformation at that temperature. The sum of the incremental quench factor values over the precipitation or transformation range is the quench factor, Q. These calculations are quickly and easily made with modern desk top computers.

The quench factor approach is usually referred to as the Avrami or 'Additivity' rule. Scheil first proposed the additivity rule to describe incubation or nucleation during phase transformations.[9] Avrami continued the analysis and showed that when the nucleation rate is proportional to the growth rate, the additivity rule is applicable.[10,11] Avrami developed different expressions to describe the rate of transformation during phase changes depending upon whether there are few or many nuclei present and depending upon the type of phase growth that occurred. Transformation rate laws were derived for transformations that nucleated on grain boundary surfaces, grain edges and grain corners.

Cahn showed that transformations that nucleate heterogeneously often obey the Avrami rule according to rate laws that could be calculated from isothermal transformation data.[12,13] Hollomon and Jaffe acknowledged the potential applicability of the Avrami principle but concluded that the agreement between calculation and experiment was imprecise for a plain carbon steel and a chromium–nickel–molybdenum steel.[14] Kirkaldy used the Avrami principle and compared the predicted start of transformation curve with the experimental start of transformation curve for nine different steels. Good agreement between the two was found in most cases, especially in view of uncertainties in the experimental curves and the assumptions made to calculate the theoretical curves.[15]

Tamura studied effects of section size and austenite grain size on the transformation rate of steel and found the additivity rule to be approximately correct during step quenching.[16] Tamura applied the additivity rule to the transformation of a eutectoid steel and was able to obtain equations describing the hardenability that gave predictions similar but slightly lower than those of Grossman.[17]

Evancho and Staley[18] used the Avrami principle to develop an expression describing precipitation kinetics in aluminium alloys. From precipitation kinetics, an expression was derived that could be used to predict strength in aluminium from the shape of a continuous cooling curve and the precipitation kinetics of the alloy

58 Quenching and Carburising

being hardened.[18-20] The general technology for heat treating aluminum alloys using quench factor concepts has been further described by Beck,[21] Vol. IV of the *9th Edition Metals Handbook*,[22] and *Aluminum Properties and Physical Metallurgy*.[23]

Generally, long times are required for carbon diffusion at high temperatures because solute supersaturation is low, and consequently the thermodynamic driving force for diffusion is low. At intermediate temperatures, however, the undercooling and thermodynamic driving force is high, and the time required to achieve a particular amount of transformation is low – typically a few seconds for carbon and low alloy steels. Pearlite nucleation usually occur first along grain boundaries because less lattice strain is necessary, and diffusion rates are higher than within grains.

At still lower temperatures, diffusion rates decrease and bainite or martensite formation begins. Martensite is the most commonly desired transformation product in quenched steels. Quenching must produce sufficiently high cooling rates at elevated temperatures to avoid ferrite and pearlite formation and retain carbon in the austenite until the martensite transformation begins.

2.2 QUENCH FACTORS

Quench factor analysis provides a single number that interrelates the cooling rate produced by a quenchant and the transformation rate of the alloy as described by the time–temperature–property (TTP) curve. The TTP curve is the mathematical representation of the start of a transformation curve that influences the property (such as hardness or strength) of interest. In practice, the quench conditions for steel must be adjusted to achieve the minimum hardness or strength needed. There is a reciprocal relationship between quench factor and both hardness and strength with lower quench factor values being associated with higher hardness and strength values. The critical value of the quench factor is the maximum value of the quench factor, Q, that will result in the desired hardness or strength and can be defined in terms of the maximum allowable amount of transformation during cooling.

2.3 QUENCH FACTOR CALCULATION

Quench factors are calculated from digital time-temperature (cooling curve) data and C_T function describing the time–temperature–property (TTP) curve which has been derived for the alloy of interest.

The TTP curve is usually described with a C_T function of the form:

$$C_T = - K_1 * K_2 * \exp\left(\frac{(K_3 * K_4^2)}{R*T(K_4-T)^2}\right) * \exp\left(\frac{(K_5)}{R*T}\right) \tag{1}$$

where:

C_T = critical time required to form a constant amount of a new phase or reduce the hardness by a specified amount (the locus of the critical time values as a function of temperature forms the TTP curve),
K_1 = constant which equals the natural logarithm of the fraction untransformed during quenching, i.e. the fraction defined by the TTP curve,
K_2 = constant related to the reciprocal of the number of nucleation sites,
K_3 = constant related to the energy required to form a nucleus,
K_4 = constant related to the solvus temperature,
K_5 = constant related to the activation energy for diffusion,
R = 8.3143 J/(K mol),
T = temperature K.

The constants K_1, K_2, K_3, K_4 and K_5 defined the shape on the TTP curve.

An incremental quench factor, q, for each time step in the cooling curve is calculated using equation (2):

$$q = \frac{\Delta t}{C_T} \qquad (2)$$

where:

q = incremental quench factor,
Δt = time step used in cooling curve data acquisition and
C_T was defined in equation (1).

The incremental quench factor, q, represents the ratio of the amount an alloy was at a particular temperature divided by the time required for transformation to begin at that temperature.

The incremental quench values are summed over the entire transformation range to produce the cumulative quench factor, Q, according to equation (3):

$$Q = \Sigma q = \sum_{T=Ar_3}^{T=M_s} \frac{\Delta t}{C_T} \qquad (3)$$

When calculating the quench factor for a particular steel, the values of T_1 and T_2 are the Ar_3 and M_s temperatures respectfully.

The cumulative quench factor reflects the heat removal characteristics of the quenchant as reflected by the cooling curve. This factor also includes section thickness effects because section thickness affects the cooling curve. Transformation kinetics of the alloy are reflected because the calculation involves the ratio of the time the metal was at a particular temperature divided by the amount of time for transformation to begin at this temperature, i.e. the position of the TTP curve in time. The calculation process is schematically illustrated in Fig. 1.

The calculated quench factor can be used to predict the as-quenched hardness in steel using the following equation:

$$P_P = P_{min} + (P_{max} - P_{min}) \exp(K_1 Q) \qquad (4)$$

60 Quenching and Carburising

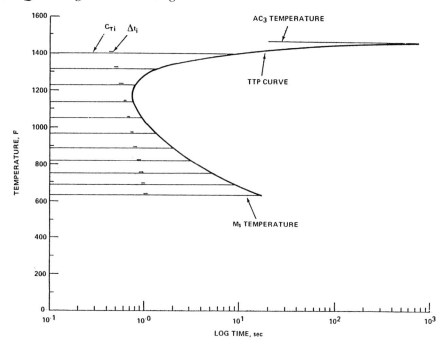

Fig. 1. Schematic illustration of the method for calculating a quench factor.

where:
P_P = predicted property,
P_{min} = minimum property for the alloy,
P_{max} = maximum property for the alloy,
exp = base of the natural logarithm,
$K_1 = \ln(0.995) = -0.00501$.
Q = quench factor.

The cumulative factor under particular quench conditions reflects the heat extraction characteristics of the quenchant as modelled by the cooling curve over the transformation range of the alloy, as well as the section thickness of the part and transformation kinetics of the alloy. An alloy with a low rate of transformation will produce a lower Q value under given cooling conditions compared to an alloy with a high transformation rate.

Quench factors calculated for different alloys might be quite different even if similar section sizes are cooled in the same quenchant because quench factors take into account individual alloy transformation kinetics by the equation describing the TTP curve for each alloy. This method of describing quench severity is quite different from Grossman numbers which are related solely to the ability of a quenchant to extract heat and not to the transformation kinetics of the alloy being heat treated.

The quench factor provided by a particular quenchant media can be experimentally determined using cylindrical, sheet or plate probes or an actual part appropriately instrumented with thermocouples, a quenchant testing apparatus in which the quenchant temperature and velocity (and concentration if an aqueous polymer quenchant is used) can be controlled and a data acquisition system for recording the temperature as a function of time. A probe or part is solution treated at the proper temperature for the alloy and quenched into a bath containing the quenchant media being evaluated at the desired velocity and temperature. Cooling curves are recorded, and the quench factors calculated from the cooling curves and the mathematical expression for the C_T function for the alloy.

2.4 APPLICATION OF QUENCHANT CHARACTERISATION BY QUENCH FACTORS

2.4.1 Hardenability of cast 4130 steel

Table 1 contains the composition and some calculated hardenability data on cast 4130 steel. The first column in the table provides composition and hardenability data on a cast 4130 steel assuming all elements to be at the low end of the composition range. The carbon range has been expanded from the normal AISI range of 0.28 to 0.33% to cover the range normally accepted in castings, i.e. 0.25 to 0.35% carbon. The second column provides data on a cast steel composition actually tested, and the third column provides data assuming all elements to be at the high end of the acceptable composition range. The composition data is followed by calculated alloy factors and ideal diameters for the low, actual and high end compositions. The calculated idea diameters (DI) ranged from 1.38 to 6.63 inches.

Table 1. Hardenability Data on Cast AISI 4130

Chemical composition	Low specification	Actual composition	High specification
	(%)		
Carbon	0.250	0.315	0.350
Manganese	0.400	0.576	0.700
Silicon	0.150	0.379	0.400
Chromium	0.800	1.140	1.500
Nickel	0.000	0.103	0.103
Molybdenum	0.150	0.244	0.250
Copper	0.000	0.125	0.125
Vanadium	0.000	0.014	0.014
Ideal diameters (in.)	1.38	4.18	6.63

Quenching and Carburising

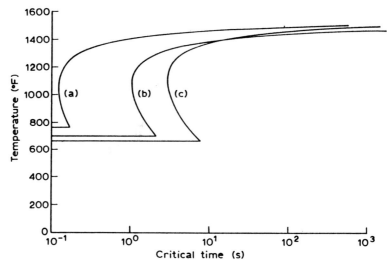

Fig. 2. TTP Curves for Cast 4130 Steel. (a) low specification composition, (b) actual composition and (c) high specification composition.

Table 2. Measured and predicted hardness at 17 positions in water quenched 4130 steel

Casting Position	Cooling Rate at 1300°F (°f/sec)	Quench Factor	Predicted Hardness (Rc)	Measured Hardness (Rc)
1	205.4	7.2	51.8	52
2	121.7	12.4	50.9	51
3	72.1	18.5	49.9	50
4	52.9	25.6	48.8	50
5	38.8	33.1	47.7	49
6	31.3	41.5	46.4	47
7	23.8	50.1	45.2	45
8	19.9	58.7	44.1	42
9	17.0	67.3	43.0	42
10	12.7	86.8	40.7	39
11	9.8	103.1	38.9	38
12	8.1	119.1	37.3	36
13	6.8	134.8	35.8	35
14	5.9	150.1	34.5	34
15	4.6	175.3	32.5	33
16	3.9	197.6	31.0	31
17	3.3	215.6	29.9	30

Most published TTT curves were developed for specific steel compositions and do not take into account the possible variations in composition and shifts in TTT curve position that can occur within the allowable composition range. Using quench factor analysis, it is possible to calculate the start of the transformation curve for a steel where a modest amount of data exists.

The approximate TTP curves have been developed for the upper and lower limit compositions as well as the actual composition of the cast 4130 steel given in Table 1. The shift in 'C curve' with composition is clearly evident in Fig. 2. These 'C curves' are plotted for times to 1000 seconds (16.7 minutes) rather than the usual 1 000 000 seconds (20 days) used in many published diagrams. Transformations that occur over a 20-day period are of little interest in heat treating operations. At 1100°F, sufficient transformation occurs in about 0.15 second in the low specification composition, 1 second in the actual composition, and after about 3 seconds in the high specification composition to cause a 1% loss in hardness.

The 'C curves' in Fig. 2 illustrate the shift in the start of transformation with alloy content, and the mathematical expression describing the curves allows

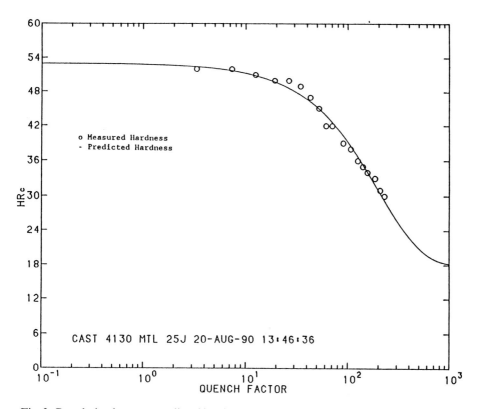

Fig. 3. Correlation between predicted hardness and measured hardness at 18 locations in a quenched part (Mtl 25).

64 Quenching and Carburising

hardness predictions to be made under a wide variety of quenching conditions. Some correlations between quench factor, predicted hardness and measured hardness in cast 4130 are presented in Table 2 and graphically illustrated in Fig. 3. The solid line in Fig. 3 represents the predicted hardness as a function of quench factor, and the data points represent measured hardness values at locations in a quenched part where cooling curves were available. These data show a good correlation between predicted and obtained hardness.

Table 3. Comparison of predicted vs. actual Rockwell (Rc) hardness values for 4140 Steel (materials 31 and 32)[a]

Material 31		Material 32	
Predicted hardness (Rc)	Actual hardness (Rc)	Predicted hardness (Rc)	Actual hardness (Rc)
57.3	56	56.5	57
57.1	56	56.3	57
56.7	55	56.0	56
56.3	54	55.6	56
56.2	52	55.6	54
55.2	53	54.8	54
51.6	53	51.6	53
53.8	50	53.6	54
49.7	49	50.0	54
48.6	49	49.0	46
44.6	42	45.5	38
25.6	31	27.5	34

[a] The elemental compositions of these steels was:

	Composition (%)	
Element	Material 31	Material 32
Carbon	0.439	0.394
Manganese	0.860	0.810
Silicon	0.259	0.268
Chromium	0.878	0.922
Nickel	0.076	0.138
Molybdenum	0.170	0.196
Copper	0.087	0.128
Vanadium	0.000	0.000
Boron	0.000	0.000

2.4.2 Hardenability of AISI 4140 steel

A similar analysis of the ability of various quench media to harden two chemistries of AISI 4140 steel was evaluated. The data in Table 3 show a good correlation between predicted and measured hardness for the two steel chemistries evaluated.

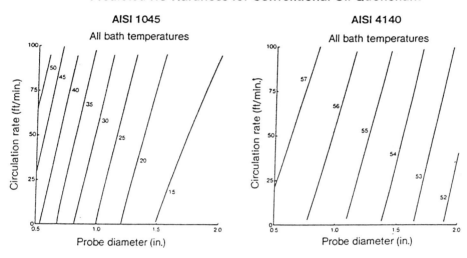

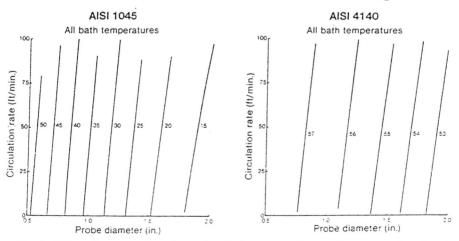

Fig. 4. Effect of cross-section size and agitation on hardness with a conventional and accelerated oil quenchant.

66 Quenching and Carburising

2.4.3 Hardenability of AISI 1045 steel

Thus far, all of the examples of use of quench factor analysis to predict hardness have been for higher hardenability steels. The use of quench factor analysis to predict the hardness for a low hardenability AISI 1045 steel was also evaluated.

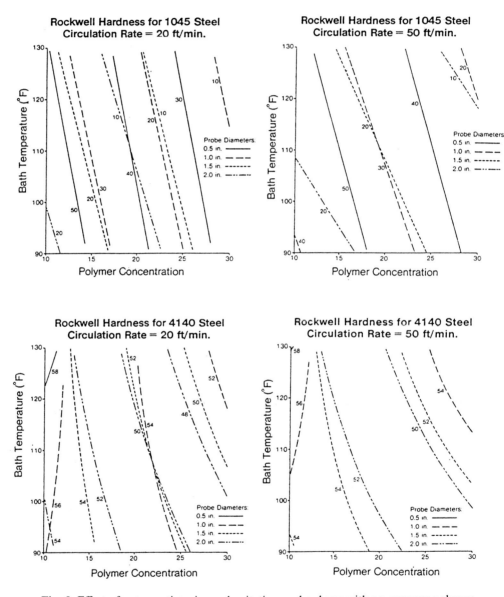

Fig. 5. Effect of cross-section size and agitation on hardness with an aqueous polymer quenchant.

The predicted and experimental hardness values agree closely until the martensite content fell below 50% at a Rc hardness of approximately 32. The reason for the inability of quench factor analysis to successfully predict hardness over the whole range of hardnesses is that it is very difficult to model the extremely rapid

Table 4. Comparison of predicted vs. actual Rockwell (Rc) hardness values for 1045 Steel (materials 27 and 28)[a]

Material 27		Material 28	
Predicted hardness (Rc)	Actual hardness (Rc)	Predicted hardness (Rc)	Actual hardness (Rc)
59.8	59	58.0	60
59.0	58	56.9	53
58.4	59	56.2	59
55.7	57	52.7	56
53.4	56	30.2	28
45.7	54	49.8	52
48.0	55	40.4	46
35.4	23	43.8	43
38.2	26	29.0	24
35.8	34	31.0	26
25.3	27	29.3	23
19.1	22	19.7	24
16.9	18	14.6	16
20.0	19	13.1	18
12.3	18	15.2	18
–	–	10.7	17

[a] The elemental compositions of these steels was:

	Composition (%)	
Element	Material 27	Material 28
Carbon	0.480	0.468
Manganese	0.834	0.749
Silicon	0.281	0.240
Chromium	0.052	0.054
Nickel	0.035	0.031
Molbdenum	0.011	0.016
Copper	0.091	0.015

transitions from softer ferritic–pearlitic structures to martensite.[25] For this reason, predicted hardness, in the range of Rc = 35–45, must be verified experimentally in low hardenability steels such as AISI 1045.

2.5 STATISTICAL MODELLING OF HARDNESS DATA

A three level statistically designed experiment was run to evaluate the effects of bath agitation and temperature on the ability of a conventional oil, an accelerated oil and an aqueous polymer solution to harden an AISI 4140 steel. For this work, linear, turbulent quenchant flow velocities of 0 (0), 50 (0.25) and 100 ft/min (0.5 m/s) around instrumented probes were evaluated. Bath temperatures of 110°F (43°C), 150°F (66°C) and 190°F (88°C) were used to bracket the most often encountered temperatures used for the oils. Probe diameters of 0.5 (1.27), 1.0 (2.54), 1.5 (3.81) and 2.0 (5.08) in. (cm) were used for this study. The predicted hardness was statistically modelled as a function of bath temperature and agitation rate. The results obtained for the oils are illustrated in Fig. 4. It was found that the as-quenched hardness of this steel was essentially independent of bath temperature over the range studied. However, the attainable hardness was dependent on quenchant velocity past the quenched part.

The results obtained with the aqueous polymer are illustrated in Fig. 5. The as-quenched hardness was found to be dependent on polymer concentration, agitation and bath temperature.

Similar cooling curve studies using quench factors and statistical analyses could be conducted with any quenchant under either laboratory or production conditions to verify performance and the effect of processing variables on the resulting hardness. With such data in hand, it will be considerably easier to establish the best quenching conditions to achieve the desired hardness.

3 SUMMARY

The function of a quenchant is to remove heat quickly enough to minimise diffusion controlled transformations to ferrite or pearlite during quenching. The quenchant must also minimise, as far as possible, thermal gradients within the part that may cause plastic deformation and residual stress. Deformation and residual stress may cause problems when the part is machined or put into service.

Quench factor analysis can be used to predict the hardness in steel from experimental and analytical cooling curves. This method permits the characterisation of a quenchant medium with a single parameter. Quench factor analysis was shown to provide a reasonably good correlation between predicted and measured hardness for higher hardenability steel alloys such as AISI 4130 and 4140 steels. Somewhat worse correlations were found with lower hardenability plain carbon steels such as AISI 1045. However, work is in progress to further refine the method for lower hardenability plain carbon steels.

The utility of the combined use of quench factor analysis to model hardness and statistical analysis to model the effect of process variables on hardening of steel using a conventional and a fast oil was demonstrated. This work showed that the performance of these oils was essentially independent of bath temperature but was dependent on agitation rate.

REFERENCES

1. M. Schwalm and H.M. Tensi: 'Heat Mass Transfer Metall'. (*Seminar Int. Cent. Heat Mass Transfer*), Sept. 1981, **45**, 563–572.
2. Th. Kunzel, H.M. Tensi and G. Wetzel: *5th Annual Congress on Heat Treatment of Materials* Vol. III, Budapest, Hungary, 20–24 October 1986.
3. H. Tensi, P. Stiltzelberger-Jacob and T. Kunzel: *Maschinemarkt, 1988*, **94**, 70–72.
4. H.M. Tensi and E. Steffen: *Steel Research, 1985*, **56**, 469–495.
5. H.M. Tensi, G. Wetzel and Th. Kunzel: 'Heat transfer', *Proceedings of Eighth International Heat Transfer Conference, San Francisco*, 3031–3035, 12–22, August 1986.
6. S.O. Segerberg: *Heat Treat., 1988*, December, 30–33.
7. P. Damay and M. Deck: Int. Heat Treating Association Meeting, Lamans, September, 1990.
8. B. Liscic and T. Filetin: *J. Heat Treating, 1988*, **57**, 115–124.
9. E. Scheil: *Arch. Eisenhuttenwes., 1935*, **12**, 565–567.
10. Melvin Avrami: 'Kinetics of Phase Change, I', *Journal of Chemical Physics, 1939*, Vol. 7, December.
11. Melvin Avrami: 'Kinetics of Phase Change, II', *Journal of Chemical Physics, 1940*, Vol. 8, February.
12. J.W. Cahn: 'The Kinetics of Grain Boundary Nucleated Reactions', *Acta Metallurgica, 1956*, Vol. 4, September.
13. J.W. Cahn: 'Transformation Kinetics During Continuous Cooling', *Acta Metallurgica, 1956*, Vol. 4, November.
14. J.H. Hollomon and L.D. Jaffe: *Ferrous Metallurgical Design*, John Wiley and Sons, New York, NY, 1947, p. 39.
15. J.S. Kirkaldy, B.A. Thomason and E.G. Beganis: 'Prediction of Multicomponent Equilibrium and Transformation for Low Alloy Steel', *Hardenability Concepts with Applications to Steel*, American Institute of Mining, Metallurgical and Petroleum Engineers, Inc., Warrendale, PA, 1978, p. 82.
16. M. Umemoto, N. Komatsubara and I. Tamura: 'Prediction of Hardenability Effects from Isothermal Transformation Kinetics', *Journal of Heat Treating, 1980*, **1** (3), June.
17. M. Umemoto, N. Nishioka and I. Tamura: 'Prediction of Hardenability from Isothermal Transformation Diagrams', *Journal of Heat Treating, 1981*, **2** (2), December 130–138.
18. J.W. Evancho and J.T. Staley: 'Kinetics of Precipitation in Aluminum Alloys During Continuous Cooling', *Metallurgical Transactions, 1974*, **5**, January 43–47.
19. *Metals Handbook*, ninth edn, 'Heat Treating of Aluminum Alloys', Vol. 4, *Heat Treating*, November 1981, pp. 675–718.
20. John E. Hatch, ed., *Aluminum Properties and Physical Metallurgy*, American Society for Metals, Metals Park, OH, 1984, 397 pp.
21. P. Archambault, J. Bouvaist, J.C. Chevrier, G. Beck: 'A Contribution to the 7075 Heat Treatment', *Materials Science and Engineering, 1980*, **43**, 1–6.
22. *Metals Handbook*, ninth edn, 'Heat Treating of Aluminum Alloys', Vol. 4, Heat Treating, November 1981, pp. 675–718.

23. John E. Hatch, ed., *Aluminum Properties and Physical Metallurgy*, American Society for Metals, Metals Park, OH, 1984, 397 pp.
24. C.E. Bates: 'Predicting Properties and Minimizing Residual Stress in Quenched Steel Parts', *Journal of Heat Treating, 1988*, **6,** 27–45.
25. J.S. Kirkaldy, McMaster University, Private Communication, 1988.

4

Use and Disposal of Quenching Media: Recent Developments with Respect to Environmental Regulations

ECKHARD H. BURGDORF

Dipl.-Ing. Karl-W. Burgdorf KG, Quenching Media and Quenching Technology,
Wielandstr. 25 D-70193 Stuttgart 1, Germany

ABSTRACT

The basic environmental hazards which occasionally arise from quenching in liquid media are the exposal of flames, fumes, vapours and toxic gases during the quenching process, waste disposal mainly being related to washing solutions and again vapours, fumes and gases during tempering. The first step to minimise these hazards is to minimise consumption through proper design of the quenching installations and proper use, control and maintenance of quenchants. In several cases, change from oil to polymer or synthetic quench oil is possible. Moreover some newly developed equipment is shown to reduce oil dragout in mass production by an automatic centrifuge system, to prevent hazardous gases during tempering of oil-quenched parts by combustion and for the recovery of marquenching salts from the washing solution through an infrared evaporation system.

1 INTRODUCTION

Because of the increasing attention to the environmental aspects of industry during the last few years, the following report presents some examples of how hazards arising from different quenching media and quenching techniques can be avoided or reduced.

2 HAZARDS CAUSED BY QUENCHING

In general, the basic use of liquid quenching media is similar for oils, water with or without additives and marquenching salts. Once the austenising time is complete,

72 Quenching and Carburising

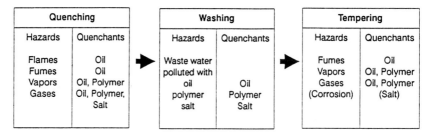

Fig. 1. Environmental hazards caused by quenching in liquid media.

the workpieces to be quenched are immersed in the quenchant. On this occasion vapours, fumes and flames are released. After the quenching process is finished, the quenched workpieces are removed from the quench tank to the next step of the process, whether it be cleaning of the parts or tempering. On the surface of the quenched parts some quenchant residues will be transferred to the washer or the tempering furnace. Of course, chemical composition and quantity of volatiles and dragout depend on the quenchant itself, but also on the following:

- shape and dimension of workpieces;
- temperature-/time cycle and atmosphere in the furnace;
- quench parameters (single/batch quenching, quenchant temperature, and concentration, quenching time, drip off time).

The main hazards arising from this are shown schematically in Fig. 1 for different steps of the process and for different quenchants. Based on this, there are different possibilities for reducing the environmental problems related to the use of quenchants. It is evident that one of the most efficient ones is to reduce consumption.

3 QUENCHING OILS

For quench oils as for all other quenchants, the most drastic method of reducing consumption is to stop quenching altogether. This has been practised in recent years in some industries with considerable success, for instance in the forging industry, where an increasing number of parts made of micro-alloyed steels is no longer quenched and tempered but hardened by controlled cooling.

Even if this is not possible – and supposedly also during the next 10 years quench oils will remain the most commonly used quenchants – there are other ways of reducing hazards such as fumes, vapour and dragout caused by the heat treatment process as can be seen in Table 1.

To give a rough idea from the figures, it should be pointed out that in an industrialised country like in Germany with a considerable heat treating industry, there is an annual consumption of quenching oils of about 10 000 tonnes. Even if there was a reduction of about 3000 tons for the quantity needed for the initial

Table 1. Methods for minimising hazards arising from quenching and their efficiency

Hazard to be reduced/avoided	Method	Efficiency (%)
Fire, smoke during quenching	Closed quench	up to 100
Fumes, vapours during quenching	Closed quench	up to 100
Fumes, vapours, gases during tempering	Reduction of dragout by increased oil temperature and extended drip off time,	10 to 20
	– washing before tempering	up to 100
	– oil removal by centrifuges	20 to 50
	– combustion of vapours	up to 100

filling of new installations, there remain approximately 7000 tonnes of real consumption.

This consumption consists – roughly estimated – of 5000 tonnes dragout and 2000 tonnes released during quenching as vapours, fumes, smoke and flames.

3.1 MINIMISING OF QUENCH OIL CONSUMPTION NEEDED FOR NEW FILLINGS

It may be supposed that most of the following basic principles for the proper use of quench oils are well known, but perhaps it should be noted that some of them considerably affect oil consumption and thus also the hazards arising from the oil quenching process.

- Take care that the volume of the quench tank is *big* enough. As a rule of thumb the volume in litres should be 6 to 10 times as much as the maximum load or the maximum throughput per hour in kg; a too small oil volume limits the lifetime of the filling.
- Choose *stable* quench oils; they are now available up to a practically unlimited lifetime.
- Limit specific heat of heating tubes in the quench tank to 1 W cm^{-2}.
- Run the quench oil at normal temperature (60–80°C).
- Restrict marquenching in oil to closed systems with protective atmosphere above the quench tank.
- Avoid excessive contact of the quench oil with air/oxygen.
- Remove scale, soot and other contaminants from time to time or continuously by decanters or filters.
- Avoid quench oil fires; be aware that extinguishing a quench oil fire with extinguishing powder or foam spoils the filling, whereas CO_2 does not affect the quench oil.

Where quench oil fillings need to be replaced frequently in spite of all the precautions noted above, which can be the case, for example, in pit type

74 *Quenching and Carburising*

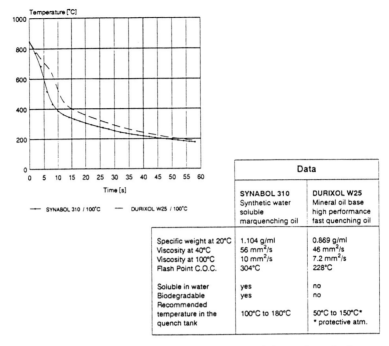

Fig. 2. Quenching characteristics (ivf Quenchotest) and data of synthetic quench oil SYNABOL 310 (BURGDORF KG).[3]

installations with extremely high throughput, the introduction of a synthetic quench oil should be considered.[3] Data and quenching characteristics of such a synthetic quench oil, recently developed for special applications are shown in Fig. 2.

The synthetic quench oil shown in Fig. 2 is based on stabilised polyethylene glycols. It does exhibit extremely high resistance to oxidation and thermal cracking and does not form asphaltenes or sludge. Also it is fully water soluble and can be removed from the quenched parts by simply rinsing. There is no need to add any detergents to the water in the washer.

A major benefit with respect to ecology is that the synthetic quench oil is fully biodegradable, i.e. the respective washing solutions can normally be disposed of in a waste water treating plant without prior separation.

3.2 REDUCTION OF DRAGOUT LOSSES IN MASS PRODUCTION

As an example of how dragout can be reduced by about 50%, Fig. 3 shows a modern centrifuge system, installed in a fully automatic heat treating line for carburising self-tapping screws after the quench and before washing.[1] Because of the unfavorable surface/weight ratio of such or similar parts, consumption will normally increase to 12 kg quench oil per tonne of quenched parts, which increases

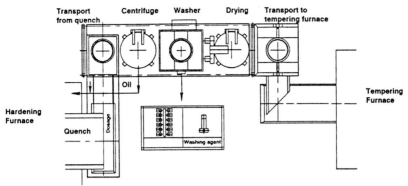

Fig. 3. Fully automatic centrifuge system for oil removal from quenched parts in mass production (WMV).[1]

not only the respective costs of replacing the dragout but also of washing and waste disposal respectively.

Oil removed from the quenched parts by means of a centrifuge can simply be returned to the quench tank; also washing becomes easier and more effective, galvanising or coating can be achieved without problems and costs for detergents, water and waste disposal are minimised.

A second method, recently suggested in a report published in the USA, is to separate the quench oil from the washing water with the aid of an oil skimmer and reuse the recovered oil for replacement of dragout losses. It has to be noted that this method is risky in so far as it is not as easy to dry oil contaminated with water and detergents as thoroughly as is necessary. It is mandatory to make sure that there is no water left in the recovered oil; more than 60% of all fires in heat treatment shops are caused by water in quench oils. And there are few things which are more hazardous than a burning oil tank.

So the use of cascades, oil skimmers, centrifuges etc. can be recommended for the removal of oil and contaminants from the washing solution, thus extending its lifetime, saving costs and providing a more effective washing process. But the recovered oil should be not used further, but disposed of or returned to the refinery for recycling.

3.3 AVOIDING HAZARDS ARISING FROM OIL VAPOURS DURING TEMPERING

There are some industries, such as manufacturers of springs and forgings, which have a large consumption of quench oils but which cannot use centrifuges to recover the quench oil adhering to the quenched parts owing to the size, shape and weight of the workpieces. Springs for the car manufacturing industry, for instance, are normally quenched directly from the forming process, which is achieved without protective atmosphere being present. So heavy scaling on the surface of the springs is normal.

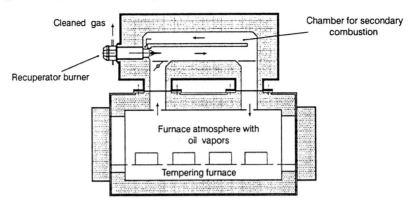

Fig. 4. Low-impact firing system for the prevention of fumes and vapors during tempering of oil-quenched parts (WS Wärmeprozesstechnik).[4]

Considerable amounts of quench oil are entrapped in the scale and, since there is no washing process between quenching and tempering, this oil gets into the tempering furnace. There it burns off or is cracked thermally, depending on temperature and oxygen present, which causes fumes and vapours.

From the technical point of view, it would be simple to avoid this, by means of a protective atmosphere during heating, forming and quenching and by washing the parts oil-free before they are put into the tempering furnace. But this would be extremely costly. A device for preventing hazardous smoke, fumes and vapours during the tempering process at reasonable cost was recently introduced to the heat treating industry in Germany by a company engaged in development and manufacturing of state-of-the-art recuperator burners. It is shown in Fig 4.

Suitable for new equipment as well as for older tempering furnaces, it consists of a sophisticated burner system mounted on top or on the side of the tempering furnace. It combusts all quench oil residues completely and thus prevents any hazardous smoke, fumes or vapours. The cleaned gas meets the requirements of German TA-Luft.

Combustion temperature is normally 900°C, which is sufficient for all mineral-based quench oils free of chlorine and chlorinated additives. If by any reason there should be chlorinated oils present, a modified system using combustion temperatures of 1200°C would be necessary and is available.

4 POLYMER QUENCHANTS

During the last 25 years, a variety of polymer quenchants has been developed, based on different polymers such as PVA (polyvinyl alcohols), PAG (normal and modified polyalkylene glycols), ACR (sodium polyacrylates), PVP (polyvinyl pyrrolidones) and PEOX (polyethyloxazoline). They are now used successfully for different applications.

For induction and flame hardening they can even be considered as standard quenching media; also in the forging industry they have been introduced successfully not only for steel grades ranging between water and oil quench, but also for replacing quench oils in several cases. A third field of application, particularly in the USA, is the quenching of aluminium alloys.

In this report it is not possible to discuss the particular properties of all polymer quenchants available. A basic common characteristic of all is that they are used for quenching not as delivered but after having been diluted with more than 50%, sometimes more than 90%, water for the ready-for-use quenching solution.

Because of their water content, there are and will continue to be some main restrictions with respect to widespread use of polymer quenchants:

- They are not suited for many of the modern atmosphere furnaces since the water vapours affect the furnace atmosphere.
- There is a tendency to non-uniform quenching results, soft pots etc. if small tightly packed components are quenched batchwise in polymer solutions.
- Close control and monitoring is mandatory.
- Risk of quench cracks and distortion is higher with polymer solutions than with quench oil.

4.1 SUBSTITUTION OF QUENCH OILS BY POLYMER QUENCHANTS

During any discussion on minimising hazards arising from the quenching process, the argument will be put forward that – with respect to health and safety protection as well as the environment – a change from oil to polymer would be preferable since the polymer quenchants can be considered as practically harmless.[5]

This is not entirely correct.

To judge the hazards arising from quenching in polymer solutions one has to look again to Fig. 1. It is true that there are no flames if one quenches in an aqueous polymer solution. But vapours do occur during quenching, as well as dragout, and hazardous vapours and gases are released during tempering if the workpieces are not washed prior to tempering.

One main advantage of polymer quenchants compared to quench oils is – apart from the fact that there are actually no open flames and no smoke during quenching – that the ready-for-use polymer solution in most cases consists not of just over 50% but of as much as approximately 90% water. And only the remaining 10%, consisting of the polymer and additives for the prevention of corrosion, foaming and fouling can possibly set free hazardous products, either during quenching or tempering.

Table 2 lists the main chemical substances which can be released, when a typical high molecular PAG polymer is cracked thermally with oxygen present.

Apparently CO_2 and water vapour are dominant, but there are also different

Table 2. Thermal decomposition of a high-molecular PAG-type polymer

Thermal and oxidative decomposition of a PAG polymer cracked thermally with oxygen present at 350°C
Gases and liquids were found as listed below

Gases	Weight (%)	Liquids	Weight (%)
Carbon dioxide	7.7	Water	17.9
Carbon monoxide	3.1	Aldehydes (various)	12.5
– Methane	0.8	Alcohols (various)	22.6
– Ethane	0.8	Esters (various)	6.5
– Propylene	7.0	Glycols (various)	9.3
		Organic acids	2.5
– Others	0.6	Others	8.7

alcohols, aldehydes, CO etc. present, which can be hazardous whenever they are released in larger quantities. Of course, the chemical nature and distribution of the decomposition products can vary widely, depending on type of polymer and additives, temperatures, atmospheres etc.

In any case, washing of the quenching parts prior to tempering and/or providing proper ventilation is recommended whenever

- highly concentrated polymer solutions of 20% and more are used;
- parts are removed from the quench tank prior to cooling down to quenchant temperature.

4.2 EXTENDING LIFETIME OF POLYMER QUENCHANTS

As stated previously, the most effective way to limit the hazards arising from the use and disposal of quenchants is to reduce consumption and to extend the lifetime of the fillings. To extend the lifetime of polymer quenchants, which meanwhile can remain in use for about 6 to 8 years in conventional installations for tank quenching, it is necessary to use clean tap water for the solution and for restoring evaporation and dragout.

Also it is necessary to provide close control and monitoring of the solution (see Table 3). In most cases, breakdown of polymer solutions is not due to ageing of the polymer but to contamination.[5] That is the reason why in induction hardening systems, where contamination by leaking hydraulic fluids, residues of metalworking fluid or rust inhibitors etc. will always be a problem, the lifetime of the polymer solution is normally limited to about 4 months. Only if parts are cleaned prior to induction hardening and the polymer solution is monitored properly, a lifetime of approximately 1 year can be achieved in such installations.

After that, the solution must be disposed of. In most cases disposal is necessary as aged oil emulsion because of entrapped oil. Disposal must and can be accepted

Table 3. Control and monitoring of polymer quenchants.[6]

Precautions	
Water used to dilute the quenching concentrate and to restore dragout and evaporation losses:	should be free of microbes, corrosive contents and the water hardness of water should be in the range of 10 to 20°dH
Testing of concentration:	daily with refractometer or areometer; additional viscosity measuring necessary for solutions used for longer periods of time
Testing of attack by microbes/disinfection:	with the aid of 'dip-slides' for selective identification of different kinds of microbes to evaluate the proper biocide
Testing of pH-value:	weekly to prevent corrosion of quenched parts and installation
Cleaning of parts to be quenched:	if possible washing of the parts prior to heat treatment, particularly before induction hardening; otherwise continuous removal of contaminants by filters or centrifuges
Circulation of polymer solution; enrichment with air/oxygen:	useful to prevent growth of anaerobic microbes whenever longer periods of non-use such as weekends, holidays etc. are expected

in that case because the quenchant volume of installations performing induction hardening is relatively low.

For larger volumes of polymer quenchants, there is an interesting method of recovering the polymer quenchant if by any reason it has changed its properties and cannot be used further. In the USA, PAG type polymers are used in large installations for quenching aluminium alloys. The workpieces are heated up in a salt bath. Therefore, salt contaminates the polymer solution which leads to an unwanted increase in quench rate and undue distortion of the quenched parts.

As shown in Fig. 5, with a heat separation process taking advantage of the fact that the PAG polymers used in quenchants separate in water and polymer at approximately 78°C, it is possible to remove the salt together with the water and to recover the polymer for making a new solution.

For other polymers, which do not exhibit the inverse solubility necessary for heat separation, a different process can be used to reach a similar result: for the recovery for instance of PVP polymers contaminated with hardening salts, an ultrafiltration process is possible, which is also applicable with PVP solutions, if their water hardness has increased to such an extent that corrosion can no longer be prevented.

80 Quenching and Carburising

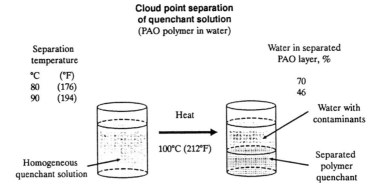

Fig. 5. Heat separation of PAG polymer quenchants.[7]

No doubt such recovery techniques are costly and only pay off for larger quenchant volumes. But they should also be taken into consideration for fillings of polymer quenchants in the forging industry which normally range between 8000 and 50 000 litres, either now or later when restrictions and costs for waste disposal are further increased.

4.3 WASTE DISPOSAL OF POLYMER QUENCHANTS

In general, the basic polymers used in the formula of polymer quenchants are relatively harmless.[5] They do not exhibit any specific toxicity. Some of them (the PVP types) are even used in medicine to replace blood plasma. But again, this is only partly true, since, as mentioned above, the polymer quenchants consist not only of polymer and water but also of additives to prevent corrosion, foaming and the growth of microbes. These additives are at present still considered non toxic or of low toxicity, except sodium nitrite which used to be part of some of the early polymer quenchants.

One also has to take into account that in most cases the polymer solution is contaminated by oils, greases, salts etc., once the time for disposal is reached.

Even if this is excluded, there remains the problem that the polymer itself has to be so stable that it is not consumed by microbes during use and that, because of this, even during the normal treatment phase in the waste water treating plant, only a low percentage of the polymer is removed by biodegradation. This means that the polymer leaves the plant practically unchanged; from here it proceeds to the next river and from there to the sea.

Because of this, it seems certain that the certificate every supplier has for his polymer quenchants, showing that they are harmless and easily disposed of, will be worthless in a few years when the total volume used in industry will be large enough to attract the attention of waste disposal authorities.

So now is the time for industry to have a look at processes such as heat

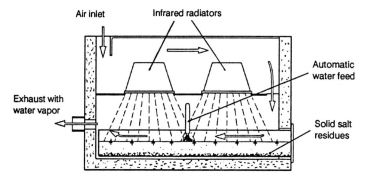

Fig. 6. Recovery of marquenching salt by an evaporation system (DEGUSSA].[8]

separation and ultrafiltration for extending the lifetime of fillings and minimising waste disposal.

5 MARQUENCHING SALTS

Because of their specific quenching characteristics marquenching salts have been used in the past mainly for tool hardening, bainitic hardening of disc springs, stampings for safety-belts and also for bearings.

Recently they have also become increasingly of interest for quenching carburised gears in some countries. Up to now there are only a few pusher type carburising furnaces run with salt quench in Germany. But this author has heard from Japan and the USA that there is a tendency to install furnaces with a separate salt quench system in recent years.

On the subject of hazards, it should be noted that marquenching salts, because of the high operating temperatures with which they are normally used, should be handled with particular care. The risks related to melting of the solid salt, water contamination etc. are well known but there is additional risk if the marquenching salt is used in a carburising furnace with an integrated quench system.

In such installations (for example in the gear manufacturing industry normally pusher type installations are used for carburizing) heavy soot deposits can form on the surface of the molten salt in the quenching vestibule due to malfunction of atmosphere regulation. As is well known, a mixture of sodium nitrite and carbon, known as 'black powder', initially invented by Bertold Schwarz some centuries ago, is explosive and thus should be avoided in a heat treating furnace.

Concerning the vapours released during the quenching process, there is a basic difference between marquenching salts and oils or polymers. In the range of temperatures used for austenising of workpieces to be quenched, i.e. approximately 780–950°C, the marquenching salts are non-volatile liquids. But it must be taken into account that a chemical reaction can occur by which NO_x gases are released during the quench process. These brownish gases, which are highly toxic,

82 *Quenching and Carburising*

are formed if the salt is overheated, particularly when a batch of small parts with an unfavourably high surface-to-weight ratio is quenched. On the other hand it should not be too difficult in workshop practice to overcome this problem by providing sufficient ventilation and reducing the batch weight if necessary.

5.1 REDUCTION OF SALT CONSUMPTION

The nitrite/nitrate salts used for marquenching are toxic. When the quenched parts are washed, which is necessary to prevent corrosion, the washing solution will have to be replaced from time to time and also it is necessary to have the waste water, which is saturated with salt, treated in a detoxification system.

A better way, recently introduced in industry by a leading supplier of salts for heat treatment, is shown in Fig. 6. With the aid of a surface evaporator system, the rinsing water enriched with salt is vaporised at 70–80°C by infrared radiation. The system operates fully automatically.

The manufacturer claims the following benefits for this system [8]:

- it is suitable for all sorts of waste water because of low operation temperature;
- in many cases reuse of the recovered salt is possible;
- no additional exhaustive air purification is necessary;
- it meets the requirements of TA-Luft (German Waste Air Regulations)
- low energy consumption

6 CONCLUSION

It has been shown that there is no general cure for the prevention of all kinds of hazards arising from quenching. But, based on the chemical and physical properties of the commonly used quenchants, there is a variety of methods for reducing considerably consumption, smoke, fumes, vapors and waste whilst improving technical results and efficiency.

REFERENCES

1. A. Müller: 'Einsatz von automatischen Zentrifugensystemen beim Behandeln von Massenteilen', *Metalloberfläche 1989*, **43**, 7.
2. William Kopke: 'Closed-loop System Solves Oil Discharging Problem and Reuses Quench Oil', *Heat Treating, 1990*, December, 25, 26.
3. E.H. Burgdorf: *SYNABOL 240, SYNABOL 310, fully synthetic water soluble quench oils*, Customers Information, Dipl.-Ing. Karl-W. Burgdorf KG, 1987.
4. G. Baur: 'Low-impact Firing System with Integrated Removal of Pollutants Using no External Energy Input', *GASWÄRME International, 1990*, **39**, (8), August.
5. S. Segerberg: 'Polymer Quenchants as an Alternative to Hardening Oils for Heat Treatment', Publication of ivf Institutet för Verkstadsteknisk Forskning, Sweden, 1985.
6. E.H. Burgdorf: *Abschreckfehler beim Induktiven Randschichthärten*, HTM Härterei-Technische Mitteilungen 44. Jhrg. 1989/3.

7. G. Totten: 'Polymer Quenchants: The Basics', *Advanced Materials & Processes* 3/90.
8. G. Wahl: '*Development and Application of Salt Baths in the Heat Treatment of Case Hardening Steels. Proceedings of ASM Heat Treating Conference 'Carburizing – Processing and Performance'*, Lakewood, Colorado, 1989, pp. 41–56.

5
Use of Fluidised Beds for Quenching in the Heat Treatment Field

RAY W. REYNOLDSON

Quality Heat Treatment Pty Ltd

1 INTRODUCTION

One of the major advantages of fluidised beds for the heat treatment of metals is the rate of heat transfer between the bed and the metal being processed.

The purpose of this paper is to examine the use of fluidised beds as a replacement quenching medium for the more conventional processes such as salt, oil and gas quenching.

In addition the use of fluidised beds to quench large tools made from hot work tool steels is examined and practical results reported.

2 PRINCIPLES IN THEORY

The theory of conventional fluidised beds has been detailed in several papers.[1,2]

The parameters controlling the cooling or quenching of metal parts in a fluidised bed are similar in most respects to heating in a fluidised bed and to a first approximation the heat transfer coefficient between an immersed surface and a gas fluidised bed can be thought to consist of three additive components:

(1) the particle convective component, a_{pc}, which is dependent upon heat transfer through particle exchange between the bulk of the bed and the region adjacent to the heat transfer surface;
(2) the interphase gas convective component, a_{conv}, by which heat transfer between particle and surface is augmented by interphase gas convective heat transfer;
(3) the radiant component of heat transfer, a_{rad}.

Thus
$$a = a_{pc} + a_{conv} + a_{rad}$$

where the approximate range of significance for a_{pc} is 40 μm – 1 mm; for $a_{conv} > 800$ μm and at higher static pressure; and for a_{rad} higher temperature (> 1000 K).

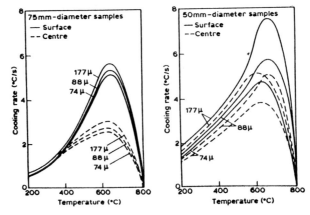

Fig. 1. The effect of alumina grain size on cooling rate curves for 75 mm and 50 mm diameter samples quenched into an ambient-temperature fluidised bed (90 MnCrV8 steel samples, 100 mm long).

In order to design a fluidised bed system which will provide the required heat transfer rates on cooling the following factors must be controlled.

2.1 PARTICLE SIZE

Tests[2,3] have been performed and reported on the effect of cooling rate with varying particle sizes between 75 and 500 μm and it has been found in practice that the optimum particle size is in the range of 100–150 μm. Figures 1 and 2 illustrate change in cooling rate with varying particle size.

2.2. TYPE OF PARTICLE AND DENSITY

A wide range of particles from glass beads, copper shot to silica sand have been tested and reported.[1,4]

The results indicate that the thermal conductivity of the particle has little or no effect on the heat transfer characteristics of the bed and that particle density is of more importance.

It has been found in practice that either silicon carbide or aluminium oxide is the most suitable and practicable particle at a density of 1761 kg cm^{-3} and that the range of optimum density is 1450–1900 kg cm^{-3}.

Figure 3 illustrates the improvement of cooling rate using silicon carbide over aluminium oxide. However, the intermixing of different particles with those used to heat the metal part is not recommended in practice and generally the increased cost of silicon carbide has precluded its use over aluminium oxide.

A literature survey indicates little work has been done on the effect of particle shape.

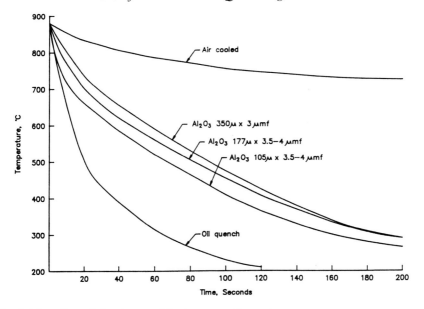

Fig. 2. The effect of alumina particle size on cooling rate.

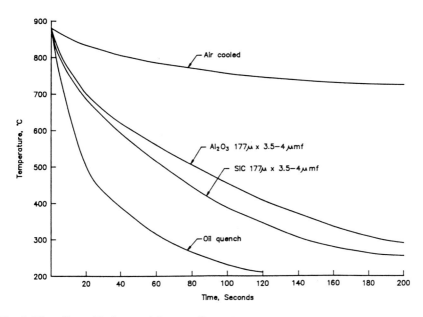

Fig. 3. The effect of bed material on cooling rate.

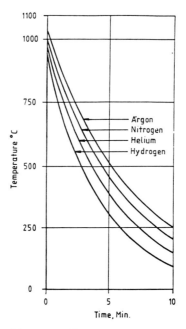

Fig. 4. Effect of gas composition on cooling rate.

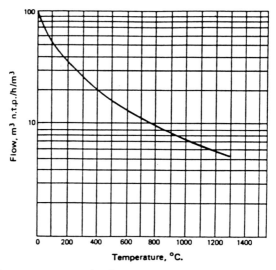

Fig. 5. Effect of temperature on the flow corresponding to minimum fluidisation for particles 0.1 mm in diameter having an apparent density of 2.

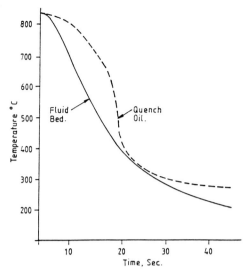

Fig. 6. Comparison of fluid bed using helium gas and oil quench rates for test specimen: 22 mm nickel sphere.

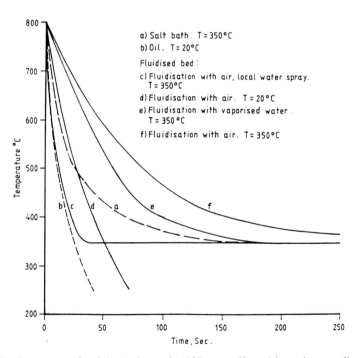

Fig. 7. Cooling curves of a plain steel sample (ϕ25 mm × 50 mm) in various cooling media measured from centre of sample.

90 Quenching and Carburising

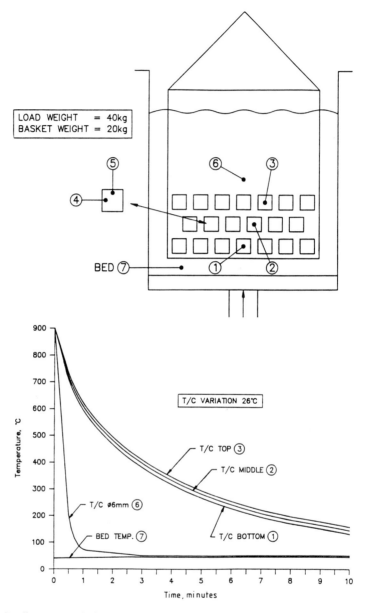

Fig. 8. Cooling rate variation in 60 mm square steel cubes from top to bottom in load of similar parts.

2.3 FLUIDISATION VELOCITY

Irrespective of the fluidising media used and the gas used to support the media, tests have shown[5] that optimum heat transfer is achieved over the general range

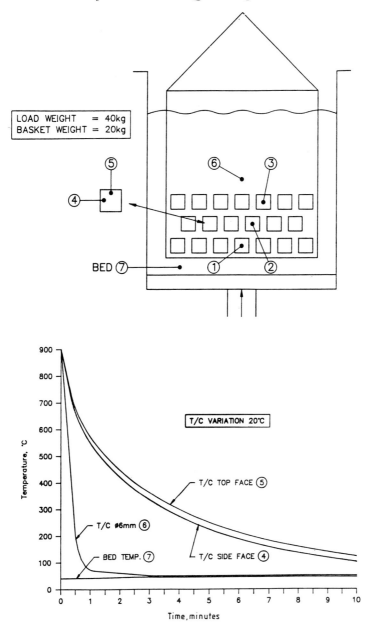

Fig. 9. Cooling rate variation in 60 mm square steel cubes from top face to side face in load of similar parts.

when the fluidisation velocity is between three and four times the minimum fluidisation velocity. These reported findings are compatible with results reported[1] elsewhere on optimised fluidisation velocity for heating in a fluidised bed where

92 Quenching and Carburising

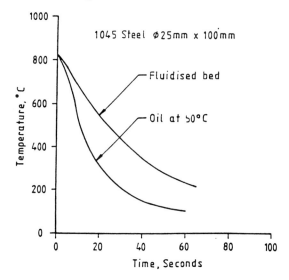

Fig. 10. Cooling curves for 1045 steel cylinders quenched in oil and fluidised beds.

two and a half to three times minimum fluidisation velocity has been found to achieve the optimum heat transfer.

The slightly increased velocity for cooling reflects the importance of the supporting gas in the convective phase component of the mechanism of heat transfer.

2.4 SUPPORTING GAS COMPOSITION

In most reported work air has been the supporting gas for the media used during cooling but as far back as the early 1970s the effect of the type of gas (i.e. thermal conductivity) on cooling rate of metal parts was illustrated as shown in Fig. 4.

The reason air has been used is that at room temperature the amount of supporting gas required to fluidise the particles is at its maximum and up to 10 times greater than gas usage at 1000°C as shown in Fig. 5. Therefore cost considerations have limited the use of other gases.

Work has been reported[6] whereby during the initial phase of cooling from hardening temperature, helium is introduced until the parts reach the martensitic start temperature at which point nitrogen is introduced to reduce the rate of cooling as well as to conserve helium.

Figure 6 indicates the claimed cooling rate comparison between oil and the fluid bed quench using helium nitrogen mixtures. The high cost of helium, between 20 to 30 times the cost per cubic metre of nitrogen, is the major problem in changing to helium and suggests that subject to satisfactory safety precautions hydrogen, a relatively cheaper gas, would further enhance the claimed cooling performance.

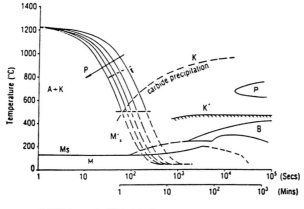
CCT diagram for M2 tool steel.

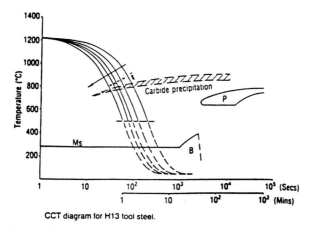
CCT diagram for H13 tool steel.

Fig. 11. CCT diagram for H13 tool steel.

2.5 MISCELLANEOUS ADDITIVES TO THE FLUIDISED BED

The use of steam or water vaporisation as an additive to the bed has been reported by Pulkkinen et al.[7] This technique necessitates that the bed temperature must be well above the vaporising temperature of water and therefore limits the application to beds operating above 200°C. It has therefore been designed to replace salt baths, for austempering of low alloy steels. Just as in salt baths the addition of water improves the cooling rate of fluidised beds as illustrated in Fig. 7.

2.6 PART CONFIGURATION AND DENSELY PACKED LOADS

The relative heat transfer of the bed versus part configuration for single parts has been well documented. Unfortunately, most tests have been performed on static

94 *Quenching and Carburising*

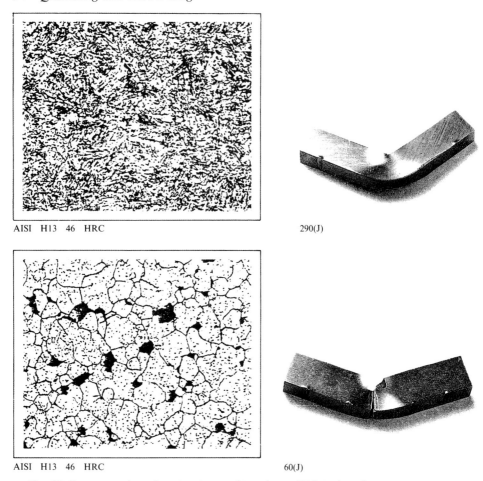

Fig. 12. Representative microstructure and toughness H13 tool steel.

test pieces and no movement to the test piece has been introduced to change or alter the less dynamic characteristics of the bed on horizontal or near horizontal surfaces, i.e. movement of the test piece horizontally, or physically improved circulation by stirring the bed as with directed agitation in oil quenching.

Similarly there is little reported work on loads packed in a similar manner to what is considered to be good practice for oil, salt or water quenching.

Figures 8 and 9 are the results of a test performed using 60 mm cubes stacked in a basket with thermocouples placed on the top face and side face of cubes in the lower middle and top layers of a basket. The results illustrate the variation that can occur over a static load.

The variation between faces, side and top of an individual part is not unusual and in oil quenching variations[5] of similar magnitudes have been shown unless adequate agitation or movement of the part is accomplished.

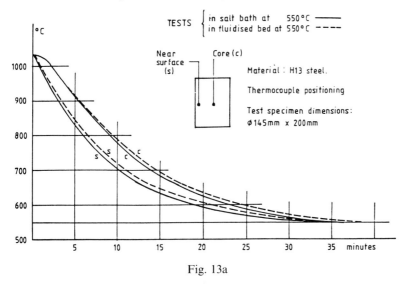

Fig. 13a

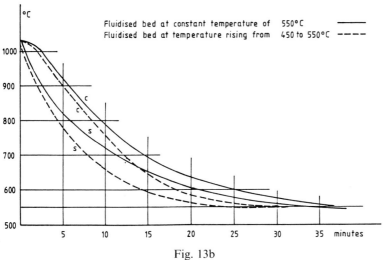

Fig. 13b

3 COMPARISON BETWEEN QUENCHING MEDIUMS

The usual quenching mediums available to the heat treating industry today can be classified as follows. Those marked with an asterisk are used to a limited extent.

Water.
Brine solutions (aqueous).*
Polymer solutions.
Oils.
Molten salts.
Molten metals, i.e. lead.*

96 Quenching and Carburising

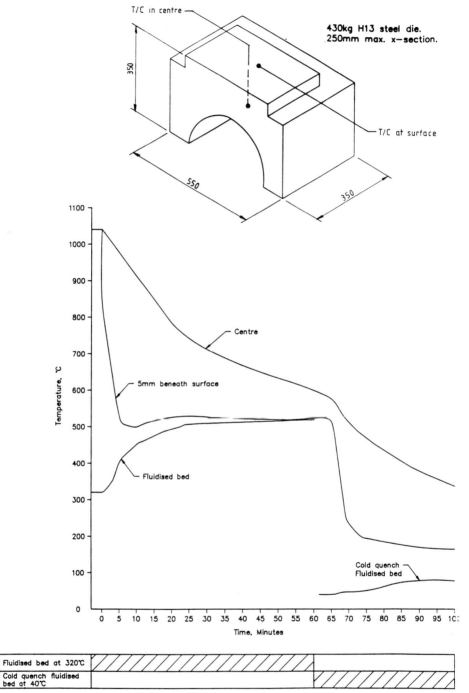

Fig. 14. Cooling curves achieved in fluidising bed quenching H13 hot work steel die casting tool.

Fig. 15. Representative H13 die casting tools.

Gases – nitrogen, argon, helium, hydrogen.
Air at pressures from 700 torr to 6 bar.
Fog quenching.*
Dry dies – water cooled.*

The basic principles involved in optimising and controlling the above quenchants are well documented and it is not the purpose of this paper to discuss these techniques but rather to compare the relevant cooling rates to fluid beds so that the fluidised bed can be assessed for areas of practical applications.

3.1 FLUIDISED BED COOLING VERSUS OIL QUENCHING

The cooling rate of a medium quench oil versus a fluidised bed using air for a 1045 steel is given in Fig. 10.

From this it can be seen that for practical purposes the cooling rate of the fluidised bed using air is not comparable to oil quenching, therefore, other than for high hardenability steels, it is not a practical alternative. However, the addition of helium does allow a possible alternative as shown in Fig. 6, and mixing of other gases is also an area for further development.

98 *Quenching and Carburising*

Fig. 16. H13 die casting tool being transferred to a fluidised quench bed after heating cycle

3.2 FLUIDISED BED COOLING VERSUS SALT BATH QUENCHING

There are two areas where the fluidised bed has been carefully examined against salt baths and these are for marquenching tool steels identified as AISI H, M, A, D and T types, and as a replacement for austempering.

3.2.1 Marquenching tool steels

Recently papers[8,9] have been published on the importance of cooling both hot work tools steels designation Type AISI H and high speed tool steels designation Type M. It has been shown that a minimum cooling rate is necessary to avoid precipitation of proeutectoid carbides at the grain boundaries. These carbides have been shown to have significantly reduced toughness of the steel in service. Therefore the cooling rate required between 900°C and 750°C is critical in minimising precipitation. Whilst this problem exists in high speed steel it is of particular importance in forging dies, extrusion tool, die casting and plastic mould tools made from the hot work AISI designation H13 steels.

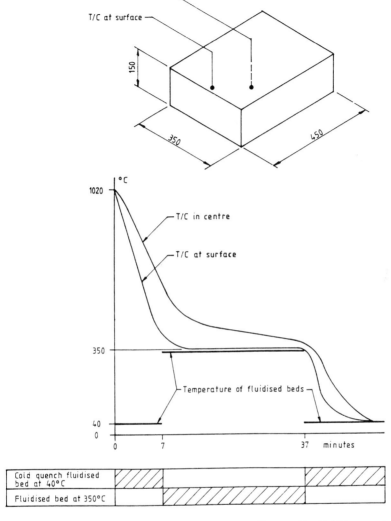

Fig. 17. Representative cooling cycle used by Liang et al[11] in cooling H13 hot work tool steels.

Figure 11 illustrates the CCT diagrams for M2 and H13 tool steels.

For example, two test pieces of H13 steel at 46 Rc are illustrated in Fig. 12. The relationship of toughness to microstructure is clearly shown.

A comparison of cooling a hot work die steel H13 in a fluidised bed and in a salt bath is given in Fig. 13 as reported by Adrien et al.[10]

To further improve the cooling rate the fluidised bed can be lowered from 450°C

100 *Quenching and Carburising*

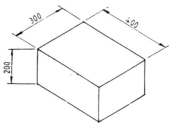

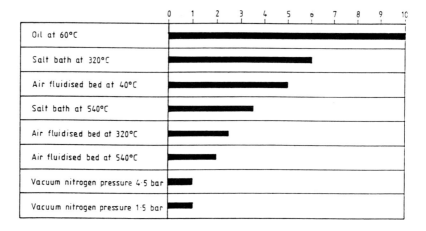

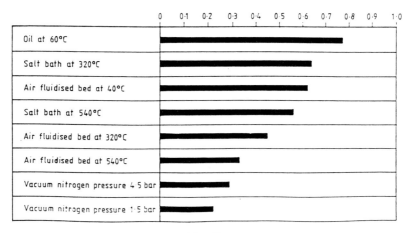

Fig. 18.

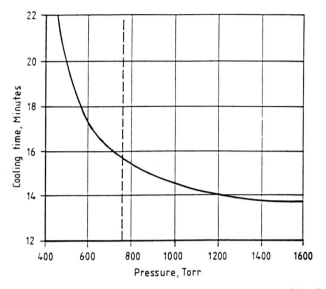

Fig. 19. Typical relationship of cooling time versus pressure quenching in a vacuum furnace.

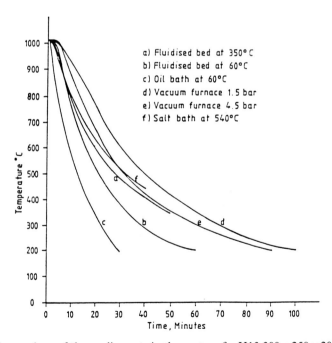

Fig. 20. Comparison of the cooling rate in the centre of a H13 300 × 250 × 200 mm steel block using oil bath, fluidised bed, vacuum furnace and salt bath.

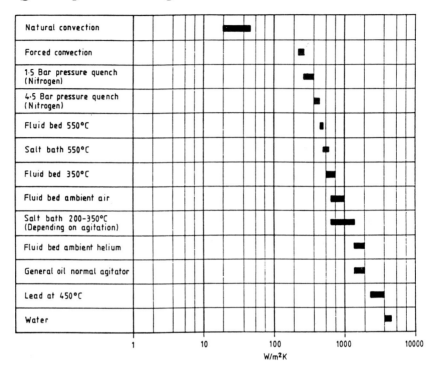

Fig. 21. Comparative heat transfer coefficients.

to 350°C which is possible with a fluidised bed and not easily replicated with the salt bath because of temperature problems.

When dies enter the bed the temperature extracted naturally increases the temperature of the bed up to 500°C–520°C whereupon the dies are removed and finally quenched in a cold quench fluidised bed operating at 50°C.

Detailed tests using this technique have been performed on large tools weighing up to 700 kg with thermocouples being placed in the centre and 5 mm beneath the surface of the tool. Typical cooling results are given in Fig. 14; obtained on dies as shown in Fig. 15; and the dies being removed from a fluidised bed just before quenching are shown in Fig. 16.

Work has been done on carefully examining the microstructure and toughness by Liang et al.[11] using a modified technique to the above where the faster cooling speed of the cold or ambient quench fluidised bed is used initially to increase the cooling rate and then, by monitoring the surface temperature, the tool is transferred to a second bed at 350°C. Typical cooling rates between 900°C and 750°C achieved in the centre of a 150 mm × 340 mm × 450 mm block weighing 227 kg are 0.75–0.6°C/second. The technique used is shown schematically in Fig. 17.

Figure 18 illustrates comparable cooling rates for different medias in the range 900°–750°C.

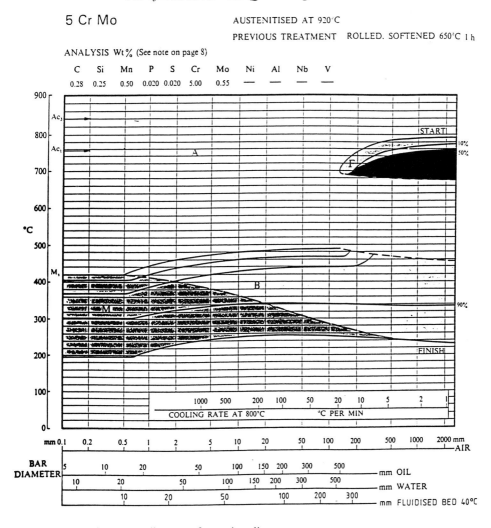

Fig. 22. Continuous cooling transformation diagram.

3.2.2 Austempering

Work has been performed by many investigators to replace the nitrate based austempering salts with fluidised beds. One such technique of using a water additive has already been described in Section 2.5.

It has been shown that without the use of such additives the cooling rate is not comparable with salt and therefore only special alloy steels or very thin sections are practicable.

The author is not aware of any practical installations using water additives in service.

3.3 FLUIDISED BED COOLING VERSUS GAS PRESSURE CONVECTION COOLING

Since the advent of vacuum furnaces the cooling of the work after the heating cycle has been a major area of research and since the early 1970s the effect of pressure quenching and the variation of gases has been examined with the result that today 4.5 and 6 bar pressure quench furnaces are available for cooling.

The effect of increasing gas pressure on cooling rate is shown in Fig. 19. In practice the difference between cooling a H13 tool with an effective cross-sectional area of 200 mm in the critical temperature area of cooling using a slight positive pressure and a 4.5 bar pressure is given in Fig. 20. In addition, the effect of cooling a similar section in a cold fluidised bed is given in Fig. 18 and compared to the 4.5 bar pressure quench furnace.

In practice the pressure quench vacuum furnace achieves slightly slower cooling rates to a fluidised bed operating at 540°C and is much slower than a cold quench fluidised bed.

Therefore in treating tools where there is a uniform section the 4.5–6 bar quench units are an alternative technique to either salt or fluidised beds. However, in treating tools of differing section size because the cooling gas is at 30°–40°C thinner sections will be cooled into the martensitic temperature range before the tool as a whole has had a chance to equalise in temperature with a resultant increase in shape distortion and the possibility of cracking.

Interrupted quench techniques in vacuum furnaces have been proposed to overcome this problem, but because of the low thermal mass of the furnace it is extremely difficult to maintain a constant temperature at the surface of the tool as can be achieved with salt or fluidised bed furnaces.

4 SUMMARY OF COMPARISON OF COOLING RATES

With the variety of quenching medias and the variation in testing and reporting between oil, pressure quenching, salt bath and fluidised bed quenching, an exact comparison is extremely difficult.

Figure 21 lists the relative range of heat transfer coefficients for the major quenching media. In addition, as a guide to the use of currently designed fluidised beds, a continuous cooling diagram for a 5% chromium steel with an additional grid for fluidised bed cooling is given in Fig. 22.

The use of the fluidised beds in practical application for the quenching of hot work tool steels is now widespread and it is estimated that in Australia alone 500 tonnes are processed each year.

Other applications of the ambient fluidised bed for cooling items such as high speed steels and specific parts after hardening and tempering are now widely accepted and used in heat treatment.

5 CONCLUSION

The use of fluidised beds for cooling and/or quenching in heat treatment is well accepted and in certain areas is now considered the most suitable method to obtain optimum results. This paper has reviewed the current literature and presented new data on the quenching of tools and dies.

ACKNOWLEDGEMENTS

I would like to acknowledge work performed by I. Gay, Heat Treatment Manager of Quality Heat Treatment on the heat treatment of the large tools and D. Liang, Technical Director of Assab Taiwan on his work on the optimum heat treatment of hot work tool steels. Both contributed significant help and information for the compilation of this paper.

REFERENCES

1. R.W. Reynoldson: 'Controlled Atmosphere Fluidised Bed for Heat Treatment of Metals', *Heat Treatment of Metals 1976*, H4 S92–100.
2. P. Sommer: *Einsatz von Wirbelschichten als Abkuhlmedium ZWF77*, 1982, 9.
3. R.W. Reynoldson: *Heat Treatment in Fluidised Beds*, 1989, ISN No. 0-7316-6274-1.
4. C.T. Moore: 'Fluidised Sand Quenching Trials for Production of Pearlitic Blackheart Malleable Iron', *BISRA Journal, 1963*, H11 S 185–190.
5. F. Kuhn: 'Die Abkukling in dir Wirbelschicht', *Warm Gas International, Sept 1980*, pp. 481–483.
6. Linde Gas Publication, 1988.
7. R. Pulkkinen *et al.*: 'Accelerated Cooling of Steels and Cast Irons in a Fluidised Bed', *4. International Congress on Heat Treatment of Metals*, 3–7 June, 1985, Berlin.
8. K. Bengtsson *et al.*: 'Heat Treatment of Hot Work Dies Using Different Techniques', *6th International Congress of the Heat Treatment of Materials*, Chicago, 1988.
9. Internal private correspondence with ASSAB Steels, 1990, based on work performed by Uddeholm, Sweden.
10. J. Adrien *et al.*: 'Trempe Etagee De Matrices en lit', *Fluidise Treatment Technique, 1976*, p 103.
11. D. Liang: Private correspondence, ASSAB Taiwan, 1991.

6
Gas Quenching with Helium in Vacuum Furnaces

BENOÎT LHOTE

L'air Liquide, Centre de Recherche Claude-Delorme, BP 126-78350 Les Loges en Josas, France

OLIVIER DELCOURT

Peugeot SA, Centre Technique de Belchamp, 25420 Voujeaucourt, France

1 INTRODUCTION

Gas quenching in vacuum furnaces has many advantages in comparison to oil or salt bath quenching. The treated parts are left clean and bright due to the purity of the quenching gas, and since there is little deformation there is no need, in most cases, for any grinding. In addition, it is a process which is more flexible to employ: rates of cooling, in particular, can be controlled and quickly adapted to each type of treatment. Finally, as there is no need for cleaning and grinding, quenching can be integrated into a mass production line.

Gas quenching is thus commonly used for the treatment of high speed tool steels, hot work or cold work steels or superalloys in aeronautics. The maximum permissible rate of cooling in a gas medium nevertheless limits gas quenching at present to metals with a moderate quenching critical rate.

The Peugeot SA Group and L'air Liquide therefore joined together to extend its application to steels with a high quenching critical rate, mass produced for case hardening treatment. Consequently, the aim of the cooperation between the two companies was to extend the performances of gas quenching.

Tests to compare the cooling properties of argon, nitrogen, helium and gas mixtures were performed on three industrial installations. Their interpretation is based on a prior, intentionally simplified study of the physical principles of the gas cooling of a load in a vacuum furnace. The presented experimental results emphasise, in addition, interactions between the nature of the gas employed and other parameters affecting quenching.

108 *Quenching and Carburising*

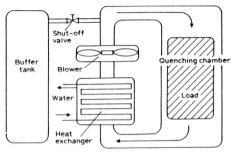

Fig. 1. Schematic diagram of gas quenching.

2 PRINCIPLE OF GAS QUENCHING

2.1 OPERATION

Vacuum furnaces equipped for the rapid gas cooling of their load operate according to one and the same scheme (Fig. 1). The furnace or quenching tank have in series with the quenching chamber a gas–liquid heat exchanger and a gas recirculating blower. Gas is first injected by way of a buffer tank connected to the chamber by a large diameter line and shut-off valve.

Before quenching, the load is at an initial uniform temperature in the vacuum chamber. The shut-off valve is opened to fill the chamber with the gas from the buffer tank up to a set pressure. Then the blower and the high flow rate of coolant in the exchanger are started. The initially cold gas then heats on contact with the load, gives off a part of its heat to the coolant in the heat exchanger before being recycled on the load by the blower.

2.2 PARAMETERS CONTROLLING COOLING

This brief description of the process highlights the main parameters controlling cooling. They can be characteristic of the treated metal, the load or even the selected process and, in particular, the gas.[1-3]

Metal. First, it is the thermal and mechanical properties of the metal which determine the treatment specifications. These properties are the critical quenching rate of the metal, its thermal diffusion, α_m, and its heat capacity C_{pm}.

Load. The cooling of the load is then related to magnitudes relative to its geometrical shape, the total quantity of heat to be removed, the surfaces available for heat transfer and the nature of the flow of gas around the parts. These are the weight, m, of the load, the thickness of the parts, their surface/volume ratio and their distribution.

Gas. The gas, which transfers the heat of the treated load to the coolant in the exchanger, is, nevertheless, the main limiting factor of the desired quenching. The action of the gas is characterised by magnitudes established by the process employed:

- its speed, which depends essentially on the performances of the blower, the driving motor and the shape of the recirculation circuit;
- its average temperature, related to the size of the exchanger;
- its pressure, adjustable as desired, and limited only by the mechanical resistance of the chamber. In practice, in France, many furnaces are designed for a maximum relative pressure of 4 bars, above which safety testing becomes more costly;
- finally, the primary factor is the nature of the gas since this determines its intrinsic transfer properties: heat conductivity λ_g, dynamic viscosity, μ_g, density, ρ_g, and heat capacity, $C_{\rho g}$.

Gases which can be used for their inert or reducing properties are argon, nitrogen, helium or hydrogen. The purpose of the study which follows is to assess theoretically the effect of their transfer properties on cooling for a given type of furnace, process and load.

2.3 MODELLING OF COOLING

2.3.1 Hypotheses

Treatment specifications determine the hardness to be obtained at different points of a mechanical part. Continuous Cooling Transformation (CCT) diagrams of treated metal enable us to relate the desired hardness to a cooling law. It is thus desirable to model the range of temperatures of a part in a load as a function of time.

This is determined by the conduction in the part and by heat exchange with the outside. Such exchanges can be broken down into:

- convection transfer between part and gas,
- radiation transfer between one part and another part, part and quenching chamber, part and gas.

A complete modelling of heat transfers requires a knowledge of the behaviour of the gas around the load, which means the speed and temperature at any point over time, as well as the temperature of other parts and the walls of the quenching chamber. Gas flow and gas temperature depend in addition on passage in the heat exchanger, the blower, and pressure drops in the recirculating circuit.

We shall limit the study to a highly simplified analytical approach to convection heat transfers between the gas and the load in the quenching chamber.

2.3.2 Solid-gas heat transfer by forced convection

The cooling of one part of a quenched load is considered here as an overall convection heat exchange between the part and the gas.

The walls of the quenching chamber, insulated and with low thermal inertia, are

110 *Quenching and Carburising*

cooled more rapidly than the load. The rate of cooling of the part is thus underestimated since radiation heat transfers from the part to the gas or towards the colder walls of the quenching chamber are neglected.

Convection transfers are considered globally between the load and the gas by determining the average temperatures for the gas (T_g) and the load (T_m). There is also an average speed of the gas, V_g, in the direction of recirculation.

When:
$$F_m = \frac{dT_m}{dt} = \text{load cooling rate,}$$

h = average convection coefficient at the gas-part interface.

$$A = \frac{S}{V_o \rho_m C_{pm}} = \text{parameter, function of shape of load and the metal}$$

where:
S = surface area of load,
V_o = volume of load,
ρ_m = density of metal

Through a balance of the enthalpy on the system making up the load, we have:
$$F_m = A\, h\, (T_m - T_g)$$

The average rate of cooling measured at a time t is a direct function of the convection coefficient h.

2.3.3 Calculation of the convection coefficient h

There have been many studies of convection heat transfer around heat exchanger tubes. Empirical formulas are used to express the convection coefficient h from dimension free numbers characteristic of flow and heat transfer. The definitions are briefly recalled below.[4]

$$\text{Reynolds number: Re} = \frac{\rho_g V_g d}{\mu_g}$$

$$\text{Nusselt number: Nu} = \frac{hd}{\lambda_g}$$

$$\text{Prandtl number: Pr} = \frac{\mu_g C_{pg}}{\lambda_g}$$

The convection coefficient is calculated with the Nusselt number, which is expressed in terms of the Reynolds and Prandtl numbers by the general formula of the form:

$$\text{Nu} = C\, \text{Re}^\alpha\, \text{Pr}^\beta$$

Where C, α, β are constants which vary with the geometry of the load and the nature of flow.

We shall select a load consisting of plain cylindrical parts, as well as two types of gas flow called F and G (Figs. 2 and 3).

The axes of the cylinders of diameter d are parallel and are arranged in a staggered, regular manner ($d = 0.1$ m; $1 = 0.2$ m). The cooling gas hits the cylinders

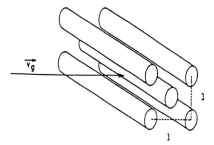

Fig. 2. Gas flow around cylindrical parts. Hypothesis F – flow perpendicular to the axes of cylinders.

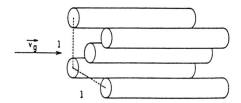

Fig. 3. Gas flow around cylindrical parts. Hypothesis G – flow parallel to the axes of cylinders.

perpendicular to their axes under hypothesis F and parallel to their axes under hypothesis G.

Grimison's formula[5] gives for hypothesis F:

$$Nu = 0.52\ Re^{0.562} Pr^{1/3} \qquad (1)$$

Under hypothesis G, flow is comparable to a flow inside a tube of equivalent diameter corresponding to the space between four cylinders. Colburn's formula[5] thus gives:

$$Nu = 0.023\ Re^{0.8}\ Pr^{1/3} \qquad (2)$$

The effect of gas speed and gas pressure on cooling is computed from expressions (1) and (2). In fact, these two parameters are involved with one and the same exponent in the Reynolds number, the density of the gas being proportional to pressure. Generally, the coefficient of convection h is proportional to the product of pressure multiplied by gas speed at a power of between 0.5 and 0.8, according to the flow hypothesis.

2.3.4 Results for different gases

Pure gases. To evaluate the effect of the nature of the gas, the coefficients of convection h for argon, nitrogen, helium and hydrogen are calculated for the following particular values:

Table 1

	Hypothesis F		Hypothesis G	
Gas	h	h/h argon (%)	h	h/h argon (%)
Ar	85	100	58	100
N2	127	149	85	147
He	215	253	88	152
H2	284	334	118	203

$$(h \text{ in W m}^{-2} \text{ K}^{-1})$$

$V_g = 10 \text{ m s}^{-1}$ $P_g = 4$ bars rel.

$T_g = 100°C$ $d = 0.1$ m

$T_m = 700°C$ $l = 0.2$ m

The resulting coefficients of convection for the two flow hypotheses are given in Table 1. Hypothesis F (flow perpendicular to the axes of the cylinders) immediately appears as being clearly more favourable to heat transfers than hypothesis G (flow parallel to the axes of the cylinders). Actual flows which occur in furnaces are nevertheless often more complex, and the ideal arrangement of parts more difficult to determine.

Other information of interest is provided by an examination of the relative behaviour of the pure gases in two extreme cases: hydrogen or helium improve resulting performances considerably in comparison to argon or nitrogen. Hydrogen involves safety problems when used in furnaces not designed for that purpose. We shall therefore concern ourselves above all with helium.

Gas mixtures. The calculation of the coefficient of convection h is extended to gas mixtures containing helium. Figures 4 and 5 illustrate results for varying concentrations of helium in argon, respectively for hypotheses F and G.

The maximum cooling performance is not necessarily obtained with pure helium: heat conductivity, dynamic viscosity and the density of the gas exert contributing effects, which serve to determine an optimum mixture for each type of flow. In the case of flow parallel to the axes of the cylinders (Fig. 5), maximum cooling is obtained with a mixture of 80% helium and 20% argon, whereas in the case of flow perpendicular to the axes of the cylinders (Fig. 4), cooling is at a maximum for pure helium.

As essentially the gas flow depends upon the type of furnace and the treated load, the gas mixture appropriate to each new treatment range can be determined. As helium is more costly than nitrogen or argon, the optimization of the gas mixture must also include process profitability criteria.

Gas Quenching with Helium in Vacuum Furnaces 113

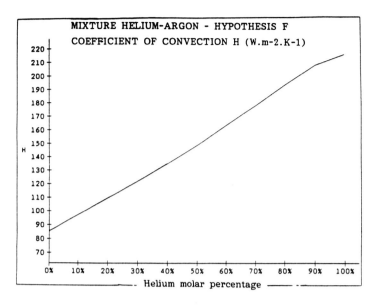

Fig. 4. Coefficient of convection *h* for helium-argon mixtures according to hypothesis F.

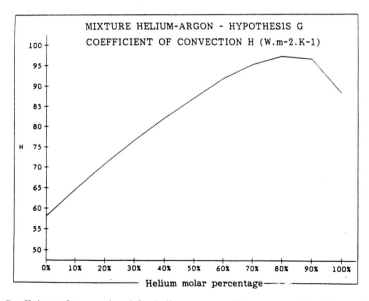

Fig. 5. Coefficient of convection *h* for helium-argon mixtures according to hypothesis G.

114 *Quenching and Carburising*

Table 2

Rel. pressure (bars)	Gas	F corner (C/min)	F corner (%)	F centre (°C/min.)	F centre (%)	Wa (kW)	Wa/ρ_g $kW \cdot m^3/kg$
3.5	Argon	1310	100	840	100	232	54
	INARC 9	1790	137	1235	147	–	–
1.5	Argon	710	100	520	100	126	54
	N2	990	139	680	131	100	56
	N2–H2	1060	149	730	140	96	56
	INARC 9	1300	183	820	158	61	65

3 EXPERIMENTAL RESULTS

The results presented now compare the cooling rates obtained in three industrial installations during normal operation with nitrogen or argon, and then with helium or a helium-based mixture. In an initial series of tests, the merits of a helium-based mixture to improve cooling are described. In the two other series of tests, we shall concern ourselves more particularly with interactions of the gas with the equipment of the furnace employed, which can be used to vary gas speed and pressure along with its nature.

3.1 QUENCHING WITH A HELIUM–ARGON MIXTURE

Several quenchings were carried out in the heat treatment shop of the SNECMA plant in Gennevilliers on parts made of a nickel alloy. The load, with a gross weight of 35 kg, is kept constant at a temperature of 1300°C, and then cooled with argon at a relative pressure of 3.5 bars in furnace with a volume of 9 m.^{3}

Table 2 shows the average cooling rates between 1300°C and 700°C for metal test parts placed at the centre (F centre) and in the corners of the load (F corner), together with the electrical consumption of the driving motor of the blower. Quenchings were carried out successively with argon, nitrogen, nitrogen and hydrogen (10%), and 70% helium and 30% argon mixture, or 'INARC 9' (INARC 9 is a registered trademark of L'air Liquide).

For quenchings carried out at the same pressure, the gases are then rated according to their cooling efficiency:

$$70\% \text{ He} + 30\% \text{ Ar} > N_2 + 10\% \text{ H}_2 > N_2 > \text{Ar}$$

The cooling rates obtained with INARC 9 at a relative pressure of 3.5 bars were reduced by water vaporisation inside the heat exchanger. The water flow rate, actually designed to remove the heat conveyed by argon, is not adequate to absorb the quantity of heat moved by the helium–argon mixture. The vaporised water thus disturbs the operation of the exchanger and minimises cooling.

Quenching at a relative pressure of 1.5 bars with INARC 9 gives metallurgical

Table 3

Gas	Rel. pressure (bars)	F top (C/min)	F top (%)	F bottom (°C/min.)	F bottom (%)	Wa (kW)
N2	2.5	99	100	67	100	57
He	2.5	131	132	98	148	7
He	4	194	196	153	228	11

Table 4

Gas	Blower speed (rpm)	F (°C/min)	F (%)	Wa (kW)
N2	3000	308	100	16
He	3000	302	98	4
N2	3300	394	128	20
He	4500	523	170	7

results which are as good as those obtained with argon at a relative pressure of 3.5 bars. In addition, and during the same time the electrical power consumed by the blower driving motor, proportional to density, was divided by 4, which eliminates a problem of excess current on startup. To be noted, by contrast, are the marked differences in cooling between the corners and the centre of the load.

3.2 HELIUM QUENCHING WITH PRESSURE VARIATION

These tests were performed at the Peugeot plant in Sochaux on case-hardened steel parts. The load, with a gross weight of 200 kg, was arranged on two levels in furnace with a volume of 2.5 m³. Quenching was carried out following a soaking temperature of 980°C.

Table 3 shows the average cooling rates between 800°C and 400°C for parts located on the lower level (F bottom) and higher level (F top). The temperature of the part was measured by a thermocouple positioned 5 mm below the surface of the metal. Helium quenching at a relative pressure of 2.5 bars gives results 30% superior to those obtained with nitrogen under identical operating conditions.

In this case, however, the relative pressure of the gas in the furnace is limited to 2.5 bars with nitrogen: at that pressure, the power consumed by the blower motor (57 kW) is already greater than its rated power (45 kW). The power transmitted to the gas by the blower being, at an identical recirculation rate, proportional to the density of the gas, it is not possible to increase nitrogen pressure without tripping

116 *Quenching and Carburising*

the blower drive motor. The use of helium lowers the power consumed by the motor to 7 kW and the pressure can be raised to the limit permissible for the furnace, or 4 relative bars. This thus gives results twice superior to those of nitrogen treatment.

In all cases, it is to be noted that cooling of the parts is not as good at the bottom level. Since the gas circulates from top to bottom, the gas is already heated by the parts on the top level when it encounters the parts on the bottom level. The cooling power of the gas is thus reduced. An increase in gas pressure in this case not only improves the coefficient of convection h, but also the uniformity of cooling in the load. Increased pressure, by increasing the quantity of recirculated gas, increases as well the total heat capacity of the gas and reduces differences in temperature between the top level and bottom level.

2.3 HELIUM QUENCHING WITH HIGHER VELOCITY BLOWING

The final tests were performed at the Peugeot Technical Centre in Belchamp with a smaller furnace (volume of 0.6 m^3), using a load with a gross weight of 40 kg and parts identical to those of the tests described above. The average cooling rates F between 800°C and 400°C are given in Table 4.

All the quenchings were performed at the maximum permissible pressure of the furnace, or 4 relative bars. Under usual operating conditions, helium in this case does not lead to any improvement in comparison to nitrogen. This result, different from those discussed in the two preceding sections, shows the decisive effect of the type of gas flow, which modifies the action on cooling of a change in gas.

At identical cooling rates, the blower drive motor is used at a much lower power with helium than with nitrogen. To take advantage of the energy still available with helium, when the maximum pressure of the chamber has been reached, we accelerate the speed of the blower. The rate at which the gas recirculates increases by as much; the power consumed by the blower varies with the cube of the gas speed. The blower was sped up to 3300 rpm for nitrogen and 4500 rpm for helium, limited by the mechanical strength of the installation.

In this instance, quenching is improved by 30% with nitrogen and by 70% with helium. Brinell hardness at a critical point of the quenched parts is 363 HB for high velocity nitrogen. It rises to 415 HB for high velocity helium quenching, with parts made of steel from the same casting as above. The studied parts, which up to now could only be quenched with oil, were able to accept correctly gas quenching with high velocity helium. The steel, used in mass production for case hardening treatment, is hardened in a satisfactory manner with helium cooling in this latter furnace.

4 CONCLUSIONS

Gas quenching, with all of its advantages, can be extended to steel or alloys with a high quenching critical rate, quenched up to now in oil or a salt bath. Performances obtained traditionally by the application of gas quenching in vacuum furnaces must nevertheless be improved.

The brief study of heat transfers during quenching demonstrates the parameters affecting cooling, as well as their interaction with the technology employed. The use of helium in particular, a gas which is a good conductor of heat, improves the performance of quenching.

Tests performed in three industrial installations show appreciable improvements in cooling when replacing the customarily used argon or nitrogen with helium. Results depend greatly on the type of furnace employed, the arrangement of the treated load and the resulting flow of gas inside the furnace. The best results are obtained by combining the use of helium with a variation of other parameters. The low level of power required for the recirculation of a gas as light as helium makes it possible, depending upon the equipment employed, to increase the pressure of the gas in the chamber or its rate of recirculation.

When replacing nitrogen with helium, one can obtain a doubling of cooling rate by raising the relative pressure from 2.5 to 4 bars in a first furnace and 70% increase in a second installation by speeding up the recirculation blower from 3000 to 4500 rpm.

The results can still be improved through the use of a helium-based mixture with better heat transfer properties. The composition of that mixture will nevertheless depend upon the type of gas flow around the treated parts. A technical and economic optimisation will determine the best mixture for each type of treatment. A more refined modelling of gas flow and heat transfer existing in the quenching chamber of a vacuum furnace is now in progress, which will establish more precisely the mixture best suited for each application.

REFERENCES

1. A. Magnee and C. Lecomte-Mertens et al: 'Modelisation du refroidissement en fours sous vide: exemple d'application aux aciers à outils', *Traitement Thermique*, 184–84 39–49.
2. W.R. Jones: 'High Velocity Gas Flow Seen as Key to Rapid Quench', *Heat Treating, Sept. 1985*, 34–35.
3. E.J. Radcliffe: 'Gas Quenching in Vacuum Furnaces – A review of Fundamentals'
4. S. Candel: *Mécanique des Fluides*, Cours de l'Ecole Centrale de Paris, 1982, pp. XV 17, 18; VIII 21–29.
5. J. Huetz: 'Thermique – Convection' p. 33, 34, 120
6. J. Friberg and J.M. Merigoux: 'Ventilateurs–soufflantes–compresseurs', *Techniques de l'Ingénieur* – B 4210 a.

NOMENCLATURE

A (m$^2 \cdot$K$\cdot$J^{-1})   Parameter characteristic of the geometry of the treated load.
C Constant
C_{pm} (J$\cdot$kg$^{-1}\cdot$K^{-1})   Heat capacity of metal

118 Quenching and Carburising

d (m)	Characteristic length of flow (cylinder diameter)
F (K·s^{-1})	Average cooling rate between 800°C and 400°C, or between 1300°C and 700°C
F_m (K·s^{-1})	Average cooling rate of metal load
h (W·m^{-2}·K^{-1})	Cefficient of convection
Nu	Nusselt number
Pr	Prandtl number
Re	Reynolds number
S (m^2)	Total surface area of treated load
t (s)	Time
T_g (K)	Average temperature of gas in quenching chamber
T_m (K)	Average temperature of metal load
V_o (m^3)	Volume of treated load
V_g (m·s^{-1})	Average speed of gas in quenching chamber
Wa (W)	Power consumed by blower driving motor
αm (m^2·s^{-1})	Diffusivity of metal
α, β	Constants
λg (W·m^{-1}·K^{-1})	Heat conductivity of gas
μ_g (kg·m^{-1}·s^{-1})	Dynamic viscosity of gas
ρg (kg·m^{-3})	Density of gas
ρm (kg·m^{-3})	Density of metal

7
Residual Stress in Quenched Spheres

D.W. BORLAND*

Department of Mechanical and Manufacturing Engineering, University of Melbourne
Parkville, Victoria, Australia 3052

and B-A. HUGAAS

BHP Melbourne Research Laboratories, 245 Wellington Road,
Mulgrave, Victoria, Australia 3170

ABSTRACT

Recent progress in the calculation of residual stress distributions in quenched steels is briefly reviewed: the finite-element method (FEM) has been developed to a high level of sophistication by including such effects as transformation plasticity and creep deformation during the heat treatment cycle. Virtually all FEM analyses use a two-dimensional calculation; the very large demands on memory space and time that are made by three-dimensional calculations have so far prevented their widespread use. The symmetry of the sphere offers the possibility of a substantial reduction in the computational complexity while still respecting the three-dimensional nature of the problem. A simplified analysis is presented that may be of sufficient accuracy for applications in which the object undergoing quenching has a sphere-like shape.

1 INTRODUCTION

The generation of residual stresses during the heat treatment of metals, especially the quenching of steel, is a subject of considerable importance because it may lead to distortion or cracking of the heat-treated part. Even if there is neither distortion nor cracking, the residual stress will add to the stress arising from service loads. In such a case the result may be beneficial, as for the fatigue resistance of parts with residual compressive stress at the surface, but it may also be detrimental.

Because of the importance of the subject it has received considerable attention and attempts to predict residual stress distributions have a long history in the literature of the heat treatment of steel.[1,2,3] Qualitative or, at best, semi-quantitative predictions, taking into account only the strains generated thermally or by the volume changes accompanying transformation, were all that could be

* Now at Department of Materials Engineering, Monash University, Victoria, Australia 3168.

attempted until the advent of high-speed digital computers. Rapid developments in computing power have since led to much more sophisticated calculations with progressively more detailed models.

Finite-element methods are now universally used for the calculation of the stress, but even the most powerful of these have still generally been restricted to two-dimensional cases. The extra demands that are placed on processing time and memory by a full three-dimensional calculation have been too discouraging, so that most available solutions are for long cylinders or for plates, where two-dimensional models are adequate.

The quenching of a spherical object represents a case where a two-dimensional approximation is clearly inappropriate. Yet the symmetry of the sphere is such that some simplification of the calculation should be possible. In this paper an approach is suggested that takes advantage of this symmetry and may offer an economical way of calculating residual stresses to an adequate degree of accuracy.

2 GENERATION OF THERMAL STRESS AND STRAIN

Thermal stress is generated whenever there is a non-uniform distribution of temperature within a body, as will be the case during all but the slowest of cooling rates. During the cooling of steel from a temperature at which it is the austenitic condition there will be further stress generated by its transformation from the face-centred cubic austenitic structure to one or more of the transformation products: ferrite, pearlite or martensite. At any appreciable cooling rate, these transformations will occur at different times in different regions and, depending on the cooling rate in relation to the hardenability of the steel, may result in different transformation products. Thus the basic problem in the prediction of thermal stress is to trace the temperature changes throughout the body as a function of time during cooling. From the temperature distribution the stresses due to thermal contraction may be determined and to these must be added the stresses arising from differences in the amount of phase transformation that has occurred in different regions.

If the response of the material to stress is purely elastic then there will be no residual stress when the part finally reaches a uniform temperature equal to that of the quenching medium, although transient stresses will exist during quenching. Residual stress at the end of the quenching process arises because these transient stresses may be high enough to cause plastic deformation, especially in regions that are still at high temperature where the yield stress is low. The calculation then involves an elastic-plastic analysis which in turn requires detailed knowledge of the plastic properties of the phases present and the dependence of these properties on temperature.

While this is a difficult enough problem in itself there are further complexities introduced by the interaction of the stress field with the course of transformation. There are two aspects of this interaction. Firstly, the presence of a stress field affects the kinetics of transformation and secondly the plastic behaviour of the material is significantly affected when the transformation occurs in the presence of an internal stress field. Furthermore, a fully realistic model should take account of viscous behaviour, i.e. creep and stress relaxation.

The development of the subject to its present state has been comprehensively and lucidly described by Fletcher.[3] There are still areas where more information is needed, notably in data relating to the properties of phases at the temperatures of relevance and in relationships between stress and transformation behaviour. The most serious deficiencies are in the ability to apply the technique to three-dimensional situations and in the calculation of residual strain, i.e. distortion, from the results of residual stress calculations.

3 THE QUENCHING OF SPHERES

The development of residual stress during the quenching of spheres is, at least in principle, a problem that should be amenable to simplifications that are not generally available when the three-dimensional problem is confronted. The symmetry of the sphere is such that the parameters of interest, the temperature and the principal stresses, must be functions only of the radius. The stress state is necessarily simple, requiring for its evaluation only the calculation of a radial stress and two equal and mutually perpendicular circumferential stresses.

The sphere may then be modelled as a series of concentric shells, the number of shells being chosen by balancing the required accuracy against the extra calculation that is required. Cooling is considered to occur by discrete steps, which may again be chosen in magnitude according to a balance between required accuracy and computational complexity. The development of stress is calculated at each decrement of temperature using a device originally proposed by Eshelby[4] for the evaluation of stresses around an elastic inclusion: consider each spherical shell as a free body, evaluate the dimensional changes that occur in it during the temperature interval, then apply suitable stresses to make it fit over the underlying shells.

The following account represents a first stage in the development and application of such a model. It uses large decrements of temperature and a primitive series of shells, in order to develop the basic calculational procedure.

4 THE PRIMITIVE MODEL

We take the sphere as consisting of two spherical shells surrounding a spherical core, the boundaries between the three elements being at $R/2$ and $3R/4$ (Fig. 1). The sphere is assumed to be initially at 850°C in a stress-free state, and to cool at a constant rate that is estimated from published data on the cooling of round bars quenched in still water, corrected to apply to the cooling of spheres by using data from the classical calculations of Carslaw and Jaeger.[5] We assume that the sphere transforms completely to martensite at the end of the quenching process.

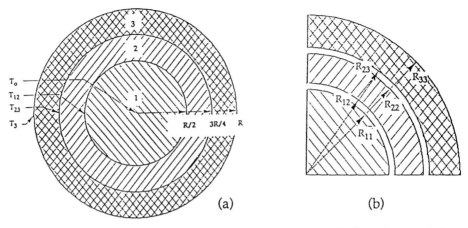

Fig. 1. The three elements of the primitive model, showing (a) dimensions, and the symbols used for temperatures; (b) the radii at which the stress is calculated.

We consider only two stages in the cooling process. The first is from the initial temperature until the surface reaches the M_s temperature. After this stage the temperature at the half-radius position will be 775°C, and we assume therefore that both shell 2 and shell 3 deform plastically during this cooling stage, so that at its end the inner and outer surfaces of each shell are at the yield stresses appropriate to the temperatures at these distances from the centre.

The second stage begins when the surface reaches the M_s temperature. We assume that no further plastic deformation of the sphere is possible. This may appear to be an unrealistic assumption but it is based on the fact that martensite is much stronger than austenite. Martensite is undoubtedly capable of plastic deformation, but even a limited amount of such deformation would lead to cracking. In practice, cracking is not observed at the surfaces of quenched spheres under a variety of quenching conditions: we conclude that it is not unreasonable to assume that plastic deformation of the surface ceases when it transforms to martensite.

If plastic deformation cannot occur at the surface then, in a sphere, plastic deformation cannot occur anywhere. This leads to a simplification of the calculation since we need consider only the initial and final states, taking into account the strains induced by thermal contraction and by expansion through phase transformation. Once again we consider the elements to behave as if they were separated and then apply suitable stresses to allow them to fit together.

5 CALCULATION USING THE MODEL
5.1 FIRST-STAGE COOLING

To carry out a calculation using the model described above we first determine the radius, R_{11}, of sphere 1 at the end of the first stage of cooling. We approximate this

Table 1. Temperatures at the end of the first stage of cooling, °C

T_0	T_{12}	T_{23}	T_3
795	775	725	250

Table 2. Stresses developed during quenching, MPa

Position	R_{11}	R_{12}		R_{22}		R_{23}		R_{33}	
Components[a]	σ	σ_r	θ_θ	σ_r	σ_θ	σ_r	σ_θ	σ_r	σ_θ
Stage 1									
Fit shell 2	−105	−105	+25	0	+130	—	—	—	—
Fit shell 3	−105	−105	−105	−105	−105	−105	+75	0	+180
Stage 2									
Fit shell 2	+165	+165	−185	0	−105	—	—	—	—
Fit shell 3	+675	+675	±675	+675	+675	+675	−1080	0	−735
Final	+630	+630	+410	+570	+595	+570	−1005	0	−555

[a] Inside sphere 1, up to R_{11}, the three principal stresses are equal. At other locations there is a radial stress, σ_r, and two equal circumferential stresses, σ_θ.

by assuming a sphere at a uniform temperature equal to the mean temperature $(T_{12}+T_0)/2$. Now, considering shell 2 as a body contracting freely as the temperature decreases and using the approximation that it is a shell at uniform temperature $(T_{12}+T_{23})/2$, we calculate the inner diameter, R_{12}.

Now, since shell 2 is at a lower temperature than sphere 1, $R_{12}<R_{11}$. Apply an internal pressure to the shell to cause it to expand (plastically) until $R_{12}=R_{11}$. The required pressure and the stresses developed in the shell are calculated using the solution given by Drucker[6] (see Appendix) in which the material of the shell is assumed to be elastic-perfectly plastic and homogeneous in its flow properties. When the two radii are equal, the shell is fitted over the sphere; the internal pressure becomes a hydrostatic compressive stress in the sphere.

The same procedure is now applied to shell 3 to make it fit over shell 2. The hydrostatic stress developed in the combined (sphere 1+shell 2) is added to the stresses previously calculated to obtain the final stress distribution at the end of this first stage.

5.2 SECOND-STAGE COOLING

Since all deformation during this stage is elastic the calculation is relatively simple. We take each of the three elements as an independent body at the appropriate

124 *Quenching and Carburising*

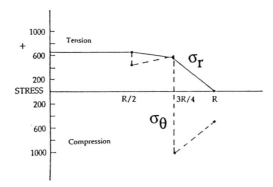

Fig. 2. The final calculated stress distribution.

temperature, which for simplicity is taken to be the mean of the inner and outer temperature of each element at the end of stage 1. Each element consists of austenite, which cools to room temperature and transforms to martensite at some stage during this cooling. The dimensional changes involved are readily calculated: because the centre starts from a higher temperature the end result is a sphere and two loose-fitting shells.

Now, to fit shell 2 over sphere 1 we imagine the application of a (negative) uniform pressure over the two surfaces to be matched. The outward displacement of sphere 1 and the inward displacement of shell 2 may readily be calculated from standard formulae of elasticity theory (see Appendix). The sum of these displacements must equal the total displacement required, and the stress fields produced in the two bodies are those resulting from the applied pressure. Shell 3 is fited over (sphere 1 + shell 2) using the same procedure and the final stress state is obtained by appropriate addition of the stress states.

6 RESULTS

In carrying out the calculations we have assumed a sphere of 90 mm diameter at the quenching temperature of 850°C. The temperatures at the end of stage 1, when the surface reaches the M_s temperature of 250°C, for the significant radii identified in Fig. 1 are given in Table 1.

Data are also required for the yield stress of austenite as a function of temperature. Comprehensive data of this kind are not readily available, and we have assumed for the purpose of this calculation that $\sigma_y = 220$ MPa at 250°C and 125 MPa at 775°C, with linear interpolation for intermediate temperatures.

The results of the calculation are presented in Table 2, which shows the stresses calculated at different stages of the calculation, and in Fig. 2, which shows the final stress distribution.

7 DISCUSSION

These results are sufficiently encouraging for the approach to be pursued further. There is an obvious difficulty with the sharp discontinuity generated during the fitting of shell 3 to the underlying (sphere 1 + shell 2) during the second stage of cooling, but this seems to be the result of using such coarse elements in subdividing the original sphere, rather than some more fundamental deficiency in the model. As the number of shells chosen for the model is increased, approximations such as the use of mean temperatures and flow stresses should become more representative of the real situation, and the computational complexity is clearly not a significant problem.

It is, however, unlikely that the procedure indicated here can be usefully extended to shapes other than spheres. It is only for the sphere that the opportunity exists for the marked simplification that is achieved because the entire sphere must behave elastically once the surface region ceases to deform plastically. This allows us, for example, to ignore the complications referred to above concerning the interaction of stress fields with transformation kinetics, since the relevant transformations occur after the surface has transformed to martensite.

REFERENCES

1. J.H. Hollomon and L.D. Jaffe: *Ferrous Metallurgical Design*, John Wiley, New York, 1947, p. 215.
2. K-E. Thelning: *Steel and its Heat Treatment*, Butterworths, London, 1975, p. 169.
3. A.J. Fletcher: *Thermal Stress and Strain Generation in Heat Treatment*, Elsevier Applied Science, London, 1989.
4. J.D. Eshelby: *Proc. Roy. Soc.* 1957, A241,376.
5. H.S. Carslaw and J.C. Jaeger: *Conduction of Heat in Solids*, Oxford University Press, 1959.
6. D.C. Drucker: *Introduction to Mechanics of Deformable Solids*, McGraw-Hill, New York, 1967, Chap. 9.

APPENDIX

The problem during the first stage of cooling is to calculate the internal pressure required to expand a thick-walled sphere under conditions where the whole of the sphere is deforming plastically. The nomenclature used by Drucker[6] in formulating the problem for a homogeneous isotropic body of tensile yield strength σ_o is shown in Fig. A.

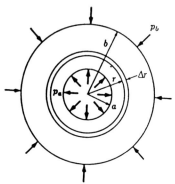

Fig. A. Drucker's nomenclature.[6]

The pressure required at $r=a$ to produce the fully plastic state, p_a^L, is given by

$$p_a^L/\sigma_o = 2 \log_e(b/a)$$

The problem in the second stage of cooling is to determine the elastic displacements in the radial direction for a sphere and a thick-walled spherical shell subjected to pressure over the external and internal surfaces respectively. For the sphere the calculation is trivial: the stress state is hydrostatic tension with $\sigma_1 = \sigma_2 = \sigma_3 = -p$, where p is the pressure at the external surface. For the spherical shell the solution is due to Lamé: using Drucker's notation the principal stresses are $\sigma_1 = \sigma_r$, $\sigma_2 = \sigma_3 = \sigma_\theta$ where

$$\sigma_r = A + B/r^3$$
$$\sigma_\theta = A - B/2r^3$$

and
$$A = (p_a a^3 - p_b b^3)/(b^3 - a^3),$$
$$B = -(p_a - p_b)/(1/a^3 - 1/b^3).$$

8

Computer Simulation of Residual Stresses During Quenching

A.K. HELLIER and M.STEFULJ

School of Materials Science and Engineering, University of New South Wales,
Kensington, NSW, Australia, 2033

M.B. McGIRR

Affiliated Consultants Pty. Ltd., Cama House, 20–22 Woodriff St, Penrith, NSW,
Australia, 2750

S.H. ALGIE

CRA Advanced Technical Development, WA Technology Park,
1 Turner Avenue, Bentley, West Australia, Australia, 6102.

ABSTRACT

This paper describes some of the principles involved in a novel computer simulation of the quenching of hardenable steel objects having simple shapes. The simulation, which incorporates processes of transient heat flow, phase transformation, elastic deformation and yielding, enables the stress distribution within the part to be evolved as quenching proceeds, and hence the final residual stresses at the end of the quench to be determined. Two different versions of the present one-dimensional simulation have been developed, one to treat hollow cylindrical steel shapes and another which is applicable to steel spheres. The former is being used to investigate the dependence of the magnitude of the surface stresses produced during quenching on the thermal history and the material properties, with the aim of combining these properties into a figure of merit, or at least producing operational quenching windows; the latter is proprietary software. Some examples of the output from this modelling procedure are included in the paper.

1 INTRODUCTION

Ideally, most heat treatments would be carried out in a high severity quench (e.g. water) resulting in maximum hardness and penetration of the section, followed by tempering to achieve the desired degree of toughness. The use of water as a quenchant has the additional advantages of cheapness, elimination of rinse cycles, and prevention of pollution caused by burn-off and oil-contaminated rinse water. Unfortunately, the use of such a severe quench medium often leads in the case of high hardenability steels to the generation of large stresses, which can result in unacceptably high levels of residual stress, distortion or quench cracking.

The processes which occur during quenching cover the entire spectrum of physical metallurgy. They include heat flow, athermal and isothermal transformations, elastic deformation and plastic flow, with fracture toughness and surface finish ultimately determining whether or not the induced stresses will be relieved by quench cracking. There is little applicable theory or quantitative design data to guide the heat treater towards a suitable quench procedure for even the simplest of part geometries. Computer simulation of the processes involved in the quenching of steel objects is therefore very attractive as a potential means of overcoming these difficulties.

A one-dimensional simulation applicable to the quenching of steel cylinders has been developed and is the subject of this paper. Such a simulation may be used in a number of ways, both as a research tool and as a means of improving industrial production. One aim of the work described in this paper is to develop a practical test which measures the propensity of steels to quench crack, and to further develop techniques for simply applying the results of such a test to the more complex part geometries encountered in practice. It is thought that such a test might be devised on the basis of the radius at which a pre-notched set of rings is found to crack.

2 CONTINUOUS COOLING TRANSFORMATION

Even in the absence of phase changes, stresses develop in a cooling body, reflecting the non-uniform density distribution produced by the temperature gradient. Plastic deformation may occur in response to these stresses, with the result that residual stresses remain after the temperature has become uniform. In hardenable steels the situation is more complex because a number of phases, with different densities, may form depending on temperature and cooling rate. This continuous cooling transformation of the austenite phase is the distinctive characteristic of the quenching of steel which must be considered in calculating the stresses which arise in the hardening process.

2.1 PHYSICAL PROPERTIES IN CONTINUOUS COOLING

The cooling of the object being quenched, the distribution of stress within it, and the occurrence of plastic flow all depend upon the bulk physical properties of the steel. These are not dependent on temperature alone, but are also functions of its microstructural state. An effective simulation of quenching must therefore take into account changes in the proportions of the microconstituents of the steel as the austenite transforms. The estimation of the microstructural state of the steel as a function of time and position in the object is thus an essential element of the modelling procedure.

Advantage can, however, be taken of this necessity to simplify the task of estimating property data. This involves the assumption that the bulk properties of the steel can be estimated by appropriately averaging the properties of the individual microconstituents. This needs some qualification in the case of structure-sensitive properties, which reflect the morphologies and distributions of the phases present. The major dependence can, however, be expressed in terms of the proportions and properties of the microconstituents.

The use of this simplification is supported by the generalisation that the essential decisions in modelling concern the inclusion or exclusion of variables. A model is an abstraction of the relevant aspects of reality constructed for specific purposes. Useful models can often be made with approximate data provided that the behaviour is correctly characterised and this can only be established by comparison with known behaviour. On this basis the simplified representation of physical data seems to be justified by experience.

The foregoing description of the principles used in the representation of physical property data does not make clear why it is essential to have this information available as a function of time and position within the object quenched. This need is particularly manifest in the treatment of latent heat, because the effects of the heat of transformation of austenite on the temperature in an object during quenching are potentially significant. Whether or not this potential is realised depends on the rate of evolution of latent heat in relation to the rate of heat transfer by conduction. This itself depends on the particular transformation product and the rate of its formation at a particular position.

Temperature measurements made in the quenching of geometrically similar objects made of steels with differing austenite transformation characteristics can show markedly different cooling histories at equivalent locations. It is quite impossible to model accurately the cooling of steels undergoing austenite transformation without taking the transformation kinetics into account.

2.2 THE QUENCHING HEAT TRANSFER COEFFICIENT

For the accurate modelling of the thermal history of steel during quenching it is necessary but not sufficient to be able to calculate, with adequate accuracy, the evolution of latent heat and the physical properties of the steel during cooling. It is also necessary to know the heat transfer coefficient at the cooling surface. A widely used measure of heat transfer in quenching is the 'severity of quench' parameter, which is defined as the ratio of the heat transfer coefficient to the thermal conductivity of the steel. The latter varies much less than the former and from the widely ranging values of the severity of quench, which characterise different quenchants and quenching arrangements, indicative values of the heat transfer coefficients can be estimated. These values do not, however, reflect the large variations in heat transfer coefficient with changing surface temperature which occur in boiling heat transfer. For quenching in water they represent the average heat transfer coefficient in the film boiling regime where the surface temperature exceeds about 400°C. They do not indicate the order-of-magnitude increase in the nucleate boiling regime around 100–400°C.

The severity of quench parameter is useful in simpler indicative estimates of the effects of quenching but it is inadequate in the accurate calculation of residual stresses. The cooling rate above 400°C is important in determining whether or not martensite forms at the surface, but transformation in the interior when the surface is below this temperature is influenced by the sudden increase in surface heat transfer when the surface falls below this temperature. Even when all phase transformations have ceased, density and mechanical property changes resulting from cooling continue to influence the stress distribution.

Accurate modelling requires accurate heat transfer coefficient data. In the case of the proprietary application, experimental heat transfer coefficient data has been determined for specific shapes in specific quenching conditions. This has been done by instrumenting the objects being quenched to determine the temperature trajectories at various internal points, observing the need to align thermocouples along isotherms within the object. In the light of the discussion above concerning the importance of the latent heat, the quenching should be carried out from a uniform temperature just below that required for austenite formation. Experimental matching of the measured and calculated thermal trajectories through adjustment of the heat transfer coefficient can then be achieved without the need to consider a varying microstructural state.

Trial and error matching of measured and calculated temperature profiles, systematically proceeding from high to low surface temperature, has been used to estimate the surface heat transfer coefficient. The accuracy of the matching should be commensurate with the accuracy and reproducibility of the measurements, which are influenced by such factors as surface scale and flow conditions in the quenchant.

2.3 KINETICS OF AUSTENITE TRANSFORMATION

Most of the physical data required to simulate the behaviour of a particular steel may be estimated from a set of microconstituent property data in which the values are functions of temperature, but are relatively independent of bulk steel composition. These data are not all readily accessible but a set which seems reasonably applicable within the overall level of approximation can be assembled. In contrast, the representation of the transformation kinetics is treated as being specific to a particular grade of steel.

Methods of calculating the transformation kinetics of austenite from relatively fundamental principles were not available for use in the present work. Such methods appear attractive because they would allow a representation which is not grade-specific. On the other hand, experimental data on the transformation kinetics of particular steels has been published and where this is not available experimental determinations can readily be made. Such determinations, which are expressed as continuous cooling transformation (CCT) data, show the progress of transformation for particular cooling trajectories. The cooling trajectory in a particular position in a component being quenched will not, however, generally correspond to one for which data are available. The problem of using CCT data is that of interpolation to arbitrary cooling trajectories.

In practice the range of trajectories is reasonably bounded by the application of the model to cooling rates representative of practical quenching and the method adopted was that applied in the widely-used *Atlas of Continuous Cooling Transformation Diagrams*.[1] This focuses on the critical importance of the time to cool through a particular temperature range immediately prior to the start of transformation. This range is arbitrary and Atkins selected a value equivalent to roughly 40°C.

There is a certain reasonableness about this idea. The transformation of austenite is controlled by cooling rate and temperature acting through the processes of nucleation and growth. A cooling time is one way of integrating these two controlling factors, the virtue of which becomes apparent when the effects of latent heat are significant. A point in the interior of the component can experience a transient rise in temperature during the quenching process, as a result of the transformation of the austenite at locations closer to the surface. It is clearly wrong to focus on the instantaneous rate of temperature change because this ignores the thermally activated time-dependent processes which have been occurring before this transient rate reversal. Yet if the cooling ceases, or is reversed, for some appreciable time, a diminution of the effects of prior cooling rate would be expected. The idea of time to traverse a critical cooling range, which can be more conveniently expressed as a rate, addresses both these considerations.

The development of internal stresses during quenching will, in general, affect the CCT behaviour of the steel. This is discussed by Thelning.[2] For the approach used here, based on experimental determinations of CCT behaviour, the absence of data, and the apparent absence of applicable theoretical calculations of the effect, have meant that this aspect of transformation is not at present taken into account.

The incorporation of data on transformation kinetics of a particular steel into the model of the quenching process is achieved by representing its CCT data in numeric form and interpolating in terms of the temperature trajectory. There is a pitfall to be avoided in doing this. The trap lies in the fact that the interpolation of CCT data provides an example of a situation, in modelling, where the precision of the representation may have to exceed the accuracy of the data.

The variation in the transformation kinetics of different samples of a steel of the same nominal grade is sufficient to require, at the least, caution in accepting a particular set of CCT data as being representative of that grade. In this sense the data cannot be considered very accurate. If, however, interpolation of this inaccurate data is not done very precisely, erroneous predictions can emerge from the model. Unrealistic differences in microstructure at points near to one another can be predicted as a consequence of errors in interpolation arising from quite small differences in temperature trajectories. These errors can be magnified by the feedback effects of the incorrect latent heat estimates which result from the errors in predicted microstructure.

Quantification of the required precision is elusive. Experience has shown that in the development of this aspect of the model, and this is probably a truth applicable to modelling in general, there is no substitute for a critical, expert and independent examination of all aspects of the calculated output to see that it passes the stringent test of conformity with reality. The goal of the computation is the stress distribution, but the microstructural prediction is a vital intermediate output.

3 MODEL FOR SIMULATION OF QUENCHING

Modelling the CCT behaviour of the steel being quenched provides the essential foundation – density and property distributions – on which the calculation of stress can be based. The complete model must combine cooling, transformation and stress generation, through a computational procedure applicable to the shape of the body being modelled. The last point is crucial. The daunting task of seeking full generality has been avoided in the present work by concentrating on the simpler cylindrical geometry. This shape is relevant to many practical situations; indeed, much heat treatment information is expressed in terms of equivalent bar shapes. More complex geometries, despite their practical importance, have not been addressed at this stage.

The cylindrical geometry, nevertheless, provides a quite complex situation. The working, and proprietary, computer program, some output of which is shown below, has been developed for the simpler, but more specific, spherical geometry.

3.1 PRINCIPLES

The model combines three essential elements:

- The first is the calculation of cooling rate. This depends on temperature, heat transfer coefficient, physical properties of the material, and the rate of latent heat evolution as the austenite phase transforms.
- The second element combines the thermal history with the CCT behaviour of the steel to calculate the phase distribution, and hence physical properties and latent heat evolution, as a function of time. This part operates interactively with the first since they are mutually dependent.
- The third element is the calculation of stress. The stress is generated by the density changes within the object which result from the temperature change. These changes have a slight effect on the dimensions of the object, and hence on the cooling rate. Effects of pressure on CCT behaviour are not considered.

Experimental CCT data is normally derived for a set of specific cooling histories. The method of generalisation to the varying temperature trajectories within the cooling object has been described in section 2.3, above. The distribution of stress is determined by applying the principle that the object will assume a state which minimises the stored elastic strain energy. These stresses may exceed the yield strength of the material, resulting in plastic deformation. At any particular time the state of the quenched object is characterised by its temperature, microstructure, thermal contraction, plastic deformation and residual stress.

The mathematical model is thus simple enough in concept but its expression in a computable form involves considerable complexity. An outline of the main steps involved is given below.

3.2 OUTLINE OF PROCEDURE

The model has been developed in terms of a sequential computation incorporating three iterative loops. The outermost loop proceeds through a sequence of step changes in time to calculate the physical properties from the microstructural state at that time. These properties are used to determine the elastic stress state. This requires the determination of the minimum of a mathematical function, for which the preferred method involves an iterative process, which is not time-related. The computed stress state will, however, be unattainable if plastic flow would occur at lower stress values. The attainable elastic stress is calculated by incorporating the third iterative loop, not time-related, in which plastic flow may alter the geometry, thus requiring a recalculation of elastic stress. This is repeated until no flow is indicated.

3.2.1 Geometrical representation

In the situation described below the quenching of a hollow cylinder is modelled. A cylindrical polar coordinate system is used, and, from considerations of symmetry, the shape may be represented by a wedge-shaped section made up of arc-shaped finite elements. The number of elements in the radial direction largely determines the accuracy of the simulation. In this form the model is one-dimensional in as much as the degrees of freedom consist only of the translation of the element boundaries in the radial direction.

3.2.2 Material properties and quench parameters

The material properties required, as functions of temperature, for the quench simulation consist of both thermal and elastic properties, namely: thermal conductivity, specific heat, density, Young's modulus, Poisson's ratio and yield stress. Also needed are the CCT data, heats of formation of the various phases and parameters describing the quench: austenitising temperature, quenchant temperature and quench severity, or, more generally, heat transfer coefficient.

3.2.3 Phase transformation

Physical properties are calculated from the mixture of microstructural constituents (austenite, ferrite, pearlite, bainite and martensite) predicted from the continuous

134 Quenching and Carburising

cooling transformation of austenite, as described above, and the properties of these individual microconstituents, expressed as functions of temperature. Each finite element is therefore assigned properties determined by averaging the values for the individual phases and microconstituents present within it, at the appropriate temperature.

3.2.4 Temperature calculation

The temperature distribution within the cylinder is calculated as a function of time using a finite difference approximation to the Fourier heat flow equation. Time is incremented in the outermost loop of the program and a forward difference transient heat flow analysis conducted within this loop. The finite difference elements are in the form of cylindrical shells with radial dimensions corresponding with the mid-points of the radial boundary positions of the finite elements used in the stress calculation (described below). The temperature at the centre of each thermal finite difference element thus corresponds to the boundary temperature in the finite element stress calculation. The temperature of the surface element approximates the surface temperature.

The rate of latent heat evolution can be a significant term in the heat balance on the representative elements. This is calculated by computing the heat of formation of the current microstructural mixture from austenite at the current temperature, and subtracting the corresponding heat of formation at the previous time step to find the latent heat evolved. This is divided by the time increment to get the required rate of heat evolution.

At each step in the computation the maximum time increment which could be used without producing instability in the solution is calculated. A smaller fraction of this value is taken as the next time step. No further temperature calculations are performed once a preselected termination temperature has been reached at a specified location.

3.2.5 Stress calculation

For the purpose of calculating stress, the steel object may be represented at any point in time as a radially symmetric elastic continuum in which the density, elastic modulus, Poisson's ratio and yield stress vary with position. The stress state at any point is given by the equations of classical elasticity. The distribution of stress is such as to minimise the elastic potential energy while maintaining the condition that at no point is the von Mises criterion for the onset of plastic strain exceeded. A finite element approach is taken.

The assembly of elements, which, through symmetry, may be represented for computational purposes by a wedge-shaped section, initially comprise the cylinder in an undistorted, stress-free condition. As cooling proceeds, the effects of the changes in density and stress state on an element could be actualised by cutting the element from the cylinder, upon which it would assume the unconfined shape which reflects its density and any plastic deformation or volumetric strain it may have undergone. The goal of the computation is the determination of the stress state which would be required to force the assembly of unconfined elements, which are distorted from their original dimensions, to once again comprise the approximation to a continuous cylinder, the external dimensions of which may vary in response to these changes. The computation therefore proceeds through the calculation of the stress required to force the distorted elements (which initially formed a wedge) back into a similar shape, i.e. with unchange sector or vertex angle, but with possibly altered radius. This must be achieved so as to minimise the strain energy while not exceeding the von Mises criterion. This is done by varying the radial positions of the element boundaries.

3.2.6 Strain energy minimisation

Each element is characterised by its unconfined dimensions and the average values of its temperature and physical properties. Strains are calculated at both radial boundaries of each element, from consideration of the actual dimensions of the element in the object with respect to its unconfined dimensions when abstracted from the object. Stresses are then calculated from these using the average elastic properties of the element together with the appropriate elasticity relationships for the element in question. The radial stress should be zero at a free surface. It should also be continuous through each radial element boundary, but these conditions are not enforced; the deviations in the computed values from the true situation may instead be used as a test of the accuracy of the computation. The stresses and strains are then used to compute the elastic strain energy for each element; these are then summed to give the total strain energy.

At each step in the strain energy minimisation for a particular temperature profile and corresponding distribution of microconstituents within the object, the corresponding equilibrium stresses are obtained. The current set of element boundary radii are first used to compute the associated strain energy for the object. The internal radii are then systematically varied by a suitable fraction of the average element dimension, and a new value of overall strain energy is calculated for each new configuration. If the latter is lower than the current minimum, then it becomes the new minimum, and the mesh is modified accordingly. This process continues iteratively with smaller and smaller radius changes until the minimum strain energy configuration is achieved to a prescribed tolerance.

3.2.7 Yielding

After energy minimisation, the stress state at both radial boundaries of each element is examined. If the von Mises stress exceeds the yield stress a suitably small increment of plastic flow is applied to change the unconfined dimensions, subject to conservation of volume (and hence mass), so as to reflect plastic flow in response to the stress state. After this change, the energy minimisation and von Mises stress checking is repeated until the criteria described above have been met. The computation then returns to allow cooling to proceed for another increment of time, after which new element properties are computed from the new microstructural state, and the new stress state is again determined.

136 *Quenching and Carburising*

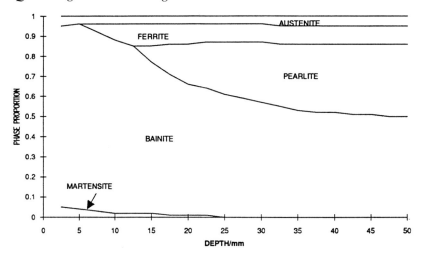

Fig. 1. Transient phase proportions during quenching of 100 mm sphere of grade 4135 steel.

3.3 OUTPUT OF THE MODEL

At present the principles and procedures outlined above have been realised, with appropriate modifications, only for spherical geometry. The outputs for this case are the temperature and microstructural state of each element, the tangential and radial stresses and strains at each radial element boundary, and the shift in radial boundary positions from the initial locations. These are manipulated externally to investigate the relationships between the variables. One of the ultimate goals is the development of simpler parameters for predicting quenching behaviour. It may be envisaged that the output of the model would be modified to express such information when it becomes available.

This prospect should be regarded with caution. The simulation gives information similar to, but in greater detail than, that which is gained from physical experimentation. The expertise required to draw conclusions from the results of physical experiments is no less necessary for the interpretation of the output of the model.

4 DISCUSSION

Examples of the results of simulated quenching are shown below. Figure 1 has been chosen to show a situation where all five microstructural constituents, austenite, ferrite, pearlite, bainite and martensite, may coexist at some positions. It shows the microstructure predicted in a 100 mm sphere of grade 4135 steel 130 seconds after the start of quenching in a medium with a constant severity of quench parameter equal to 6 ft^{-1} (corresponding to agitated oil).

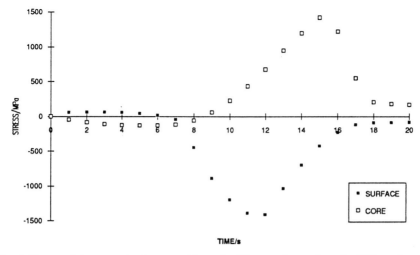

Fig. 2. Tangential stresses in the quenching of a 25 mm sphere of grade 4135 steel.

The stresses shown as functions of time in Fig. 2 are for a 25 mm sphere of grade 4135 quenched in a medium with severity of quench equal to 18 ft^{-1} (corresponding to agitated water). The rapid reversals of stress, from a situation with tensile stresses at the surface balanced by compressive stresses in the core, to a state of high internal tension with a compressed surface, results from the formation of martensite, starting at the surface. The high calculated internal tensile stresses may not be sustainable; the model does not at this time incorporate a criterion for tensile failure. Plastic deformation is, however, treated as an integral part of the calculation.

Experience to date, exemplified by these illustrations, is that the model and its expression in a computer program for spherical geometry does work in as much as its predictions can be rationalised. External verification through measurement of actual surface residual stresses is proceeding.

Even without the stress analysis, modelling the cooling while taking CCT data into account is capable of giving useful insights into quenching. The sequence of transformations in the austenite is a case in point. The formation of pearlite is associated with relatively slow cooling rates while fast cooling, which occurs near the surface, may produce martensite. While the following possibility may be well-enough known, it is interesting that people who can claim considerable experience have been surprised to see how transformation may well be complete in the hot interior while the much cooler surface is still austenite. Another misconception is to necessarily link the relatively slow cooling rates associated with pearlite formation to slow transformation. The time taken to complete the transformation from austenite to pearlite may be quite short even at slow cooling rates, and the microstructure of the resultant pearlite may be very fine. In such cases the effects of the high rate of latent heat evolution are quite significant. These examples have been obtained by modelling specific situations; they should not be taken as general truths applying to all steels.

Finally, it should be noted that the quantitative results are quite dependent on the physical property data for the individual microconstituents of the steel. In particular, the yield stress values set limits on the residual stress; the residual tangential surface stress cannot exceed the yield stress of the surface element. The significance of the lack of accurate data on individual phases is something that must be judged in the light of experience gained with the modelling process.

5 CONCLUSIONS

A procedure for simulating the processes involved in the quenching of hollow steel cylinders has been developed, and the main steps involved in the modelling procedure have been outlined. This procedure has been realised in a computer program for the more specific case of spherical geometry, and some typical examples of the output obtainable have been presented.

Modelling has its most significant applications in studying the effects of changes in variables whose interactions are sufficiently complex to produce results which are not obvious. As the results obtained with the model to date have shown, on this basis, the quenching of steel is a valid subject for modelling.

REFERENCES

1. M. Atkins: *Atlas of Continuous Cooling Transformation Diagrams for Engineering Steels*, American Society for Metals, Metals Park, Ohio, USA, 1980, pp. 225–227.
2. K-E. Thelning: *Steel and its Heat Treatment*, Butterworths, London, 2nd edn, 1984, pp. 197–204.

9

A Mathematical Model to Simulate the Thermomechanical Processing of Steel

P.D. HODGSON, K.M. BROWNE, D.C. COLLINSON, T.T. PHAM and R.K. GIBBS

BHP Research—Melbourne Laboratories, 245 Wellington Road, Mulgrave, Victoria, Australia, 3170

ABSTRACT

An integrated mathematical model has been developed to follow the thermal and microstructural evolution during the hot deformation and continuous cooling of steel. The thermal model is based on the finite-difference method with appropriate boundary conditions for the forming and cooling operations. The surface heat transfer coefficient for spray cooling is adequately represented using an equation based on the water flux and surface temperature of the steel. The microstructure evolution during multistage deformation and cooling is modelled using equations for each of the microstructural events occurring. The transformation model during continuous cooling is explicitly incorporated within the thermal model due to the change in thermophysical properties and the heat generated during each of the transformation reactions.

1 INTRODUCTION

Modern steel producers are increasingly using control of the entire thermal and deformation history after solidification to tailor the final mechanical properties of the product. The use of the *processing route* to achieve particular properties with less reliance on the base composition of the steel has been given the general description of thermomechanical treatment (TMT) and thermomechanical controlled processing (TMCP). It is also often shown that the application of TMCP technologies can lead to greater uniformity of properties over a batch of products. While these technologies were initially developed for plate steels, they have since been applied, in a more limited fashion, to strip, bar, rod and even forgings.

The final microstructures produced by TMCP range from simple ferrite pearlite structures through to complex multiphase microstructures. The microstructure depends on the steel composition, the state of the austenite at the point of

140 *Quenching and Carburising*

transformation and the thermal history during cooling to room temperature. Therefore, accurate control of the final microstructure, and hence properties, requires either extensive practical experience or the use of physical or mathematical models. With the increasing cost of plant trials and the limited range of compositions and processing routes that can be investigated from a single series of trials, there is now increasing interest in the application of simulation techniques.

The physical simulation of various TMCP processes can be performed in the laboratory using experimental rolling mills and cooling facilities, or other more controllable methods such as hot torsion or compression. In the authors' laboratories both rolling[1] and torsion[2] have been used in the investigation of TMCP processing routes for rolling and forging. It has often been found, though, that these are best used in conjunction with appropriate mathematical models if the results are to be scaled from the laboratory to the production environment. There is the further potential for models to *control* the final microstructure dynamically, although this is not practised at present.

The models described in the following sections can be divided into three types:

(1) thermal and microstructural models for the deformation phase;
(2) thermal and microstructural models for post deformation cooling;
(3) structure-property relationships.

In each case there is a certain degree of interaction between the microstructural events occurring and the resultant thermal and mechanical history. Of major importance in the quenching of steel is the heat generated during transformation and the change in thermophysical properties with the amount of each phase and the temperature. Hence, it is necessary to integrate the microstructural models into the basic thermal and mechanical models if the TMCP process is to be accurately controlled.

Each of the above submodels will be discussed in the following sections, with particular emphasis on the application to hot rolling, although it should be noted that similar approaches can be taken for forging, or even reheat and controlled cooled steels.

2 DEFORMATION MODELS

2.1 THERMAL MODEL

The solution of the differential equation describing heat transfer within the workpiece has been predominantly solved at these laboratories using the finite difference technique. The choice of this technique has been based on the requirements of fairly rapid run times on small PCs combined with easy integration with other code for the microstructure models. Various finite difference schemes have been used (e.g. explicit 1 and 2-D, implicit 1-D and alternating direction implicit 2-D), with the choice depending on the nature of the process. For example, for large objects that have reasonably long interdeformation times (>2 s) and relatively slow cooling rates, the explicit schemes offer the most efficient solution. However, these schemes are not unconditionally stable and may not be suitable for thin products. More details regarding this are given elsewhere.[3-6] There has rarely been a need to consider the finite element methods, except for a limited range of problems where greater accuracy is required for the roll-gap region.

The boundary conditions during the interdeformation periods consist of cooling by convection and radiation to the atmosphere. In the explicit model it is possible to model these separately using the equations:

$$q_c = h_c (T_s - T_a) \quad (1)$$

$$q_r = \varepsilon\sigma(T_s^4 - T_a^4) \quad (2)$$

where:
q_c = the convective heat loss (W m^{-2})
q_r = the radiation heat loss (W m^{-2})
h_c = the convective heat transfer coefficient (W m^{-2} °C)
T_s = the surface temperature (K)
T_a = the ambient temperature (K)
ε = the emissivity
σ = Boltmann's constant (W m^{-2}K^4)

However, in the implicit schemes it is necessary to combine the two terms to allow inversion of the temperature matrix. Hence a combined heat transfer coefficient, h, is used:

$$h = h_c + \varepsilon\sigma (T_s^2 + T_a^2)(T_s + T_a) \quad (3)$$

The combined heat transfer coefficient is then calculated at each time step for every surface node. Forced cooling is handled in a similar manner as described in a following section.

During the deformation period there is heat loss from the surface by conduction into the cooler rolls or platens, while heat is generated throughout the material due to the work of deformation. Examples of these calculations for flat[3,4] and shaped[5,6] rolling are described elsewhere. A comparison of the measured (using imbedded centreline thermocouples) and the predicted temperature using a 1-D model with the above boundary conditions during laboratory plate rolling is shown in Fig. 1.

2.2. MICROSTRUCTURE EVOLUTION MODEL

The development of mathematical models to predict the austenite microstructure and resultant room temperature microstructure for multistage hot forming operations was originally proposed by Sellars and coworkers at Sheffield University in the late 1970s.[7,8] Using this approach, models are obtained for each of the individual microstructural events that occur during and after deformation and these are then integrated into the relevant process model(s) for that forming operation.

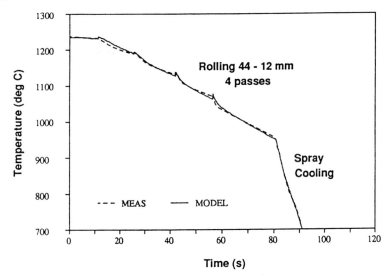

Fig. 1. Comparison of measured and predicted temperature history during experimental rolling.

The current authors have developed similar models (summarised in Table 1) to describe the following microstructural events for C-Mn and microalloyed steels[9,10]:

- static recrystallisation kinetics and recrystallised grain size,
- dynamically recrystallised grain size,
- metadynamic recrystallisation kinetics and resultant grain size,
- grain growth kinetics following complete static and metadynamic recrystallisation.

Figure 2 shows an example of the predicted austenite grain size during multipass rolling on a plate mill. In this case the grain size is refined after each pass by static recrystallisation, and then undergoes coarsening during the interdeformation period, and after the final deformation, by normal grain growth. It has been found for C-Mn steels deformed at temperatures above 1000°C that the grain size at the start of transformation is almost entirely determined by the grain growth kinetics, while the degree of deformation plays little, or no, role. Therefore, the only effective method of grain size refinement for these high temperature forming processes is to add precipitate forming elements which lead to grain boundary pinning.

AlN is effective for high N containing steels, or low reheating temperatures. However, at the typical N levels of 30 to 60 ppm in modern steels, grain coarsening will occur once the steel is reheated above 1100°C. The addition of small amounts of Ti leads to the formation of very stable Ti(C,N) precipitates which limit the grain size during reheating to less than 20 μm and to less than 30 μm after deformation.[11] The effect of a small Ti addition to a C-Mn steel is also shown in Fig. 2 for a plate rolling schedule. The control of austenite grain size by small

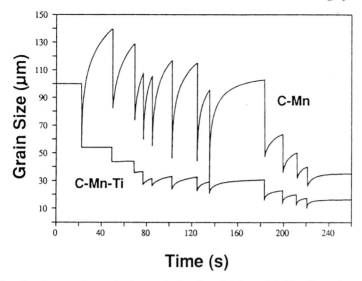

Fig. 2. Predicted austenite grain size evolution for C-Mn and C-Mn-Ti steels.

additions of Ti is the basis of recrystallisation controlled rolling (RCR)[11] and recrystallisation controlled forging (RCF).[11,12]

In rod and bar rolling the situation becomes slightly more complex. Here we still have high finishing temperatures, but there are often quench units located a very short time ($\ll 1$ s) after the final pass. Recent work[13] has suggested that the austenite grain size here is determined by the metadynamically recrystallised grain size. This is a function of strain rate and temperature, only. It is expected that metadynamic recrystallisation will also be of importance in forging where dynamic recrystallisation has been initiated.

3 MODELS FOR POST DEFORMATION COOLING

As for the thermal models during deformation, these models consist of a numerical solution for the basic conduction equation combined with equations that represent the boundary conditions. Again the finite different method has been preferred. However, some of the approximations which may be suitable for the relatively slow cooling during forming are often not adequate for forced water cooling.

If we consider an example of the cooling of a large plate, where longitudinal and transverse heat transfer effects can be ignored, then the basic conduction equation is:

$$\frac{\delta}{\delta x} k \cdot \frac{\delta T}{\delta x} + \dot{Q} = \rho \cdot c \cdot \frac{\delta T}{\delta t} \qquad (4)$$

Table 1. Equations and constants for metallurgical model

	Equation		C-Mn(+V)	C-Mn-Ti	C-Mn-Nb
Static recrystallisation	$X = 1 - \exp\left[-0.693\left(\dfrac{t}{t_{0.5}}\right)^n\right]$	(1)	$n=1$		$n=1$
	$t_{0.5} = t_0 \varepsilon^{-p} d_0^q Z^r \exp\left(\dfrac{Q_{rex}}{RT}\right)$	(2)	$t_0 = 3.6 \times 10^{12}$ $p = 2.5$ $q = 1$ $r = 0$ $Q_{rex} = 230$ kJ/mol $Q_{def} = 300$ kJ/mol	as for C-Mn	$t_0 = f(Nb)$ $p = 1.5$ $q = 1$ $r = 0$ $Q_{rex} = 250$ kJ/mol
Zener–Hollomon parameter	$Z = \dot{\varepsilon} \exp\left(\dfrac{Q_{def}}{RT}\right)$				
Metadynamic recrystallisation	$X = 1 - \exp\left[-0.693\left(\dfrac{t}{t_{0.5}}\right)^{1.5}\right]$	(3)	$k_{md} = 0.53$ $n_{md} = -0.8$	as for C-Mn	as for C-Mn
	$t_{0.5} = k_{md} Z^{n_{md}} \exp\left(\dfrac{Q_{md}}{RT}\right)$	(4)	$Q_{md} = 230$ kJ/mol	as for C-Mn	as for C-Mn

Recrystallised grain size – Static	$d_{rex} = A\varepsilon^{-a}d_o^b$	(5)	$A = 0.9 - 1.1$ $a = 0.67$ $b = 0.67$	as for C-Mn
Dynamic	$d_{dyn} = kZ^{-0.23}$	(6)	$k = 1.6 \times 10^4$	as for C-Mn
Metadynamic	$d_{md} = k'Z^{-0.23}$	(7)	$k' = 2.6 \times 10^4$	as for C-Mn
Grain growth	$d^m = d_o^m + kt\exp\left(\dfrac{Q_g}{RT}\right)$	(8)	$m = 7$ $k = 1.5 \times 10^{27}$ $Q_g = -400$ kJ/mol	$m = 10$ $k = 2.6 \times 10^{28}$ $Q_g = -420$ kJ/mol	$m = 4.5$ $k = 4.1 \times 10^{23}$ $Q_g = -435$ kJ/mol
Ferrite grain size	$d\alpha_o = (\beta_0 + \beta_1 C_{eq}) + (\beta_2 + \beta_3 C_{eq})\dot{T}^{-0.5}$ $\quad + \beta_4(1 - \exp(\beta_5 d\gamma))$ (9) $d\alpha = d\alpha_o(1 - 0.45\sqrt{\varepsilon_r})$ (10)		$C_{eq} < 0.35$ $\beta_0 = -0.4$ $\beta_1 = 6.37$ $\beta_2 = 24.2$ $\beta_3 = -59.0$ $\beta_4 = 22.0$ $\beta_5 = 0.015$	$C_{eq} > 0.35$ $\beta_0 = 22.6$ $\beta_1 = -57.0$ $\beta_2 = 3$ $\beta_3 = 0$ $\beta_4 = 22.0$ $\beta_5 = 0.015$	$d\alpha = d\alpha_{lim} + A(1.6 - \varepsilon_r) + B\dot{T}^{-0.5}$ (11) $A = f(d\gamma)$ $B = f(\varepsilon_r)$

where:

$\dot{Q}$ = the rate of volumetric heat generation (W m^{-3}),
c = the specific heat (J/kg°C),
k = the thermal conductivity (W/m°C),
ρ = the density (kg m^{-3}),
x = the displacement (m).

In the thermal models described above, for the deformation phase, it is assumed that the thermal conductivity is not a function of position and hence this equation reduces to:

$$k \cdot \frac{\delta^2 T}{\delta x^2} + \dot{Q} = \rho \cdot c \frac{\delta T}{\delta t} \tag{5}$$

However, where there are steep temperature gradients and, particularly, where the amount of a given phase is a function of position, it is not possible to assume that the variation of k with position can be ignored. Therefore, the finite difference models for quenching were derived using the complete expansion of equation (4).

To further improve the accuracy of the finite difference solution, without incurring processing penalties, the mesh is often modified so that there are more nodes located close to the surface with a larger spacing between the interior nodes. In the case of round or cylindrical product the nodes have been based on equal areas surrounding each node. This produces a decreasing node spacing when moving from the centre to the surface of the bar.

Equations (4) and (5) also contain a term for the volumetric rate of internal heat generation. For continuous cooling after deformation there is heat evolved during the diffusional phase transformations, with the heat generated by the pearlite reaction usually leading to an overall increase in temperature for typical cooling rates in medium to high carbon steels. However, rather than solve equations (4) or (5) with this term explicitly incorporated into the finite difference scheme, the approach used is to assume that the transformation effects can be decoupled. In this way the rate of heat generation is assumed to be zero during the solution of the differential equation. Any transformation and associated heat of transformation that takes place during that time interval is estimated separately, as explained in a later section, and this temperature rise is then added to each node, prior to the next time step where the process is repeated.

In all models described in this paper the thermophysical properties (c, k and ρ) are determined separately for each node as a function of the temperature at that node and the volume fraction of each phase at that node.

3.1 HEAT TRANSFER IN WATER COOLING

When water is applied to a hot steel surface, the rate of heat transfer is controlled by the thermal properties of the steel, the water and the interface. Heat flow within the steel is largely determined by its thermal diffusivity ($k/\rho c$), while heat flow into the water is by conduction and convection. The interface flow is controlled by the surface total heat transfer coefficient, designated h, and the difference in

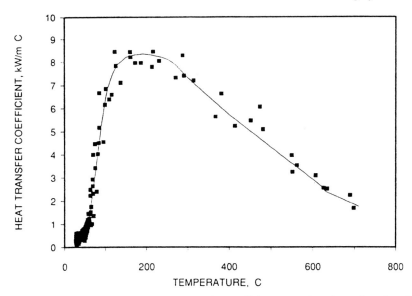

Fig. 3. Experimentally measured heat transfer coefficient, as a function of steel surface temperature, for a 60 × 40 × 0.6 mm stainless steel coupon heated to between 700°C and 950°C and vertically quenched into still water.

temperature between the steel surface and the applied water. The total heat transfer coefficient includes both convective and radiative components, as in the case of natural air cooling previously, and is defined as:

$$h = \frac{q}{T_s - T_w} \quad (\text{W m}^{-2} \text{K}^{-1}) \qquad (6)$$

where q is the total heat flux at the surface (W m^{-2}) and T_s and T_w (°C) are the surface and water temperatures, respectively.

The current methods of water cooling steel include immersion quenching into still or stirred water, high velocity water immersion tubes, hydraulic sprays, air–water mist sprays, laminar flow nozzles and water curtains. The heat transfer coefficient for water cooling is a function of the method used, the steel surface temperature and the rate at which the water is applied or flowing over the surface. In general, though, the heat transfer coefficient has the same type of surface temperature dependence for all methods of water cooling. The heat transfer coefficient for quenching into still water is typical; measured values are shown in Fig. 3.

At surface temperatures close to the water temperature, the heat transfer from the steel to the water is regulated by the conductivity of the water. As the steel temperature increases above that of the water, natural convection increases the heat transfer. The natural convective currents and heat transfer coefficient vary with the steel mass, shape and orientation. In general, though, the heat transfer coefficient increases rapidly with increasing temperature until a steam barrier at the interface begins to dominate the heat flow and reduces the heat transfer coefficient with increasing temperature until the film is fully established at about 700°C. At higher temperatures the heat transfer coefficient tends towards a fairly constant value. These effects can be different in different types of cooling systems; for example, moving water has a very much greater convective component than still water. The variations in heat transfer coefficient with surface temperature, water flow and system characteristics are most important in modelling the temperature history of the water cooled steel and hence its microstructure.

3.1.1 Hydraulic spray cooling

The heat transfer coefficient associated with hydraulic water sprays has a similar type of dependence on surface temperature as immersion quenching. The other important variable in this case is the water flux rate. While there are some data in the literature for water spray cooling, none cover the full temperature and water flux rates used in steel-making and fabrication processes. Accordingly, the convective heat transfer coefficient h_c, associated with hydraulic spray cooling of steel, has been determined from experimental data measured in these laboratories and combined with published data.

The most reliable published heat transfer coefficient for water cooling surfaces in the range 400–800°C is that of Mitsutsuka,[14] which is based on seven independent data sources and is also in excellent agreement with more recent work by Ohnishi et al.[15] Mitsutsuka's function for the convective component of the heat transfer coefficient is:

$$h_c = 3.31 \times 10^6 \, \dot{w}^{0.616} \, T_s^{-2.445} \qquad (7)$$

where $\dot{w}$ is the water flux rate and T_s is the surface temperature. In this temperature range the heat transfer coefficient decreases with increasing temperature. At the higher temperatures, somewhere above 600°C, it is well established[14,16–19] that the heat transfer coefficient becomes less temperature dependent although there is no universal agreement on its values at high temperatures. On the other hand, at low temperatures, it has been shown[16,18,20,21] that the heat transfer coefficient has a maximum somewhere between 200°C and 300°C. To accommodate both low and high temperature behaviour, Mitsutsuka's function was modified to include a maximum at low temperatures and temperature independent values at high temperatures, by employing a function of the form:

$$h_c = A \cdot \dot{w}^B \cdot [T_s - n(T_s - T_n)]^c \cdot m \qquad (8)$$

where m and n are parameters which modify the function at T_m and T_n:

$$m = 1 - \frac{1}{\exp\{(T_s - T_m)/M\} + 1} \qquad (9a)$$

$$n = 1 - \frac{1}{\exp\{T_s - T_n)/N\} + 1} \qquad (9b)$$

A, B, C, M and N are constants while T_m and T_n determine the temperature of the maximum and the temperature above which h_c becomes constant, respectively.

To determine the above constants, laboratory measurements were made by recording the temperature drop in a 100 × 50 × 16 mm type 304 stainless steel block when cooled from initial temperatures between 700°C and 900°C. The block was transported horizontally, with the largest faces vertical, through two spray headers for accurately measured times between 0.5 and 2 seconds. Each spray header was 50 mm × 450 mm in area and contained a regular pattern of 100 2 mm holes. Water fluxes of 36 and 62 Ls^{-1} m^{-2} were used. Heat loss was calculated from the average temperature loss of the whole specimen determined by extrapolating the slow temperature changes before and after quenching to the start and finishing times of spraying. This method avoids the difficulties of making dynamic temperature measurements which are always influenced by the thermal inertia of the thermocouples. The estimated radiation loss was subtracted from the total heat loss.

The values of the constants were determined by maximising the regression coefficient relating the experimental and calculated temperature drop (using a 1D finite difference model) with respect to these constants. Initially, Mitsutsuka's values of A, B and C were used and M, N, T_m and T_n were determined. A, B and C were then re-determined. The re-determined values of B and C were not significantly different to Misutsuka's and did not warrant changing. The final equation was then:

$$h_c = 3.15 \times 10^6 \dot{w}^{0.616} \left[700 + \frac{T_s - 700}{\exp\{(T_s - 700)/10\} + 1} \right]^{-2.455}$$

$$1 - \left[\frac{1}{\exp\{(T_s - 250)/40\} + 1} \right] \quad (10)$$

This equation is valid for surface temperatures from 150°C to at least 900°C and for water flux from 0.16 to at least 62 Ls^{-1} m^{-2}. Figure 4 shows h_c as a function of temperature for several different water flux rates. The correlation between the calculated temperature drops during experimental cooling using the new equation and those actually measured is shown in Fig. 5.

3.1.2 Cooling at high water flux

Because of the characteristic temperature dependence of the heat transfer coefficient, high water flux rates at the surface of a block of steel quench the surface rapidly to a fairly low temperature around 200°C and this then remains fairly constant during the quenching process. The internal cooling of the steel is then dependent on its thermal diffusivity rather than the water flux. The calculated cooling of a 12 mm thick slab by water fluxes of 50 and 100 Ls^{-1} m^{-2}, shown in Fig. 6, illustrate the typical quenching behaviour at high water flux rates and indicates the small effect of doubling the already high water flux.

This characteristic of water cooling has two important practical effects. Firstly, the water quenching of steel at very high water flux rates is essentially regulated by quenching time and, hence, is greatly affected by processing speed. Secondly, very

150 *Quenching and Carburising*

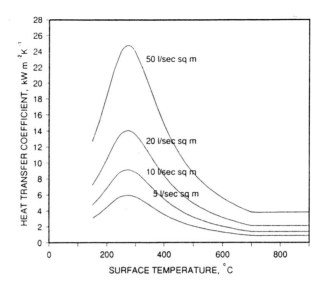

Fig. 4. The variation of the heat transfer coefficient for water spray cooling of steel as a function of steel surface temperature and water flux.

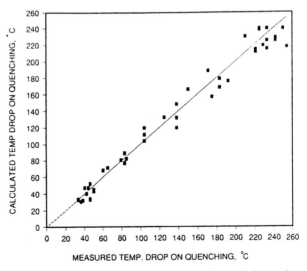

Fig. 5. Comparison of the calculated temperature drop of a steel plate using equation (10) with the experimentally measured temperature.

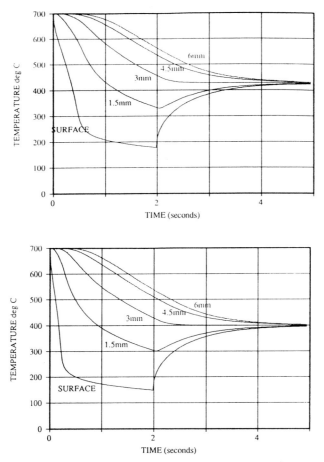

Fig. 6. The temperature distribution in a 12 mm thick slab, cooled from 800°C on both sides by a water spray flux of 50 Ls^{-1}m^{-2} (a) and 100 Ls^{-1}m^{-2} (b). Note that doubling the water flux only reduces the final temperature by a further 25°C.

efficient cooling is possible (in terms of water use) by rapidly quenching the steel surface to about 200°C with high water flux, then decreasing the water flux while maintaining the surface at a temperature where the heat transfer is a maximum.

3.2 MICROSTRUCTURE MODELS DURING TRANSFORMATION

3.2.1 Transformation kinetics

The transformation model is currently restricted to the formation of ferrite, pearlite and martensite with no account taken of the various bainites that can form during high rate cooling of certain steels. The approach taken is similar to that proposed by a number of authors,[22-24] where the progress of transformation for the diffusion reactions is followed by an Avrami type equation, assuming additivity, while the fraction of martensite formed is modelled as a function of holding temperature below the martensite start temperature.

For the diffusional phase transformations of austenite to ferrite and pearlite, the formation of a new phase on cooling is only possible once the temperature is below the equilibrium transformation temperature. This temperature is dependant on the alloy content of the steel. The calculation of the equilibrium transformation temperature for austenite to ferrite, the Ae_3, is based on the method developed by Kirkaldy and Baganis[25] and has been extended to include the elements vanadium and niobium. This technique calculates and compares the chemical potential of the austenite and ferrite phases at a given temperature for all of the elements present in the steel. The equilibrium transformation temperature is obtained when the two chemical potentials are equal. While this method is complex mathematically, it offers the advantage of being able to handle a wide range of steels. Other empirical approaches to obtain the Ae_3 temperature tend to be limited to a particular steel or group of steels.

However, for the austenite to pearlite reaction, the equilibrium transformation temperature (Ae_1) is an empirical equation by Andrews.[26] Andrews regressed this equation from data on 196 steels that was obtained from a wide range of sources. The equation obtained was:

$$Ae_1 = 723 - 10.7 \text{ Mn} - 16.9 \text{ Ni} + 29.1 \text{ Si} + 16.9 \text{ Cr} + 6.38 \text{ W} \tag{11}$$

The use of the empirical formulation for the Ae_1 has been justified by the fact that the Ae_1 temperature does not vary greatly with composition in contrast to the Ae_3.

As the temperature is reduced below the equilibrium transformation temperature for the steel there is an amount of time needed before an observable amount of new phase forms. This incubation time is dependent on the thermal history of the steel below the equilibrium transformation temperature.

Many groups have attempted to model the incubation period using the Scheil law[27] and assuming additivity during both the incubation and transformation periods. Recently Hawbolt and coworkers[22,23] have shown that additivity does not hold for the incubation period in eutectoid and hypoeutectoid steels, although it does hold during the growth period. The current model is based on their work. In this model the incubation time is based on the observed transformation start temperature from continuous cooling experiments, and is formulated as a function of undercooling and the chemistry of the steel.

Once the incubation process is completed the growth of new phase begins. The kinetics of the growth of ferrite and pearlite are described using the Avrami equation.[28-30] This equation has the following form:

$$X = 1 - \exp(-b(T)\, t^n) \tag{12}$$

where:

X = the volume fraction transformed,
t = time,
b, n are constants for a particular steel at a given temperature.

According to the Avrami equation the amount of new phase formed will eventually reach unity at long holding times. However, in the case of ferrite formation there is an equilibrium volume fraction of ferrite that cannot be exceeded at a given temperature. For this reason the equilibrium amount of ferrite which could be formed is calculated at each time step and then used to normalise the Avrami equation. Thus, equation (12) is modified to give:

$$X = X_{eqm}[1 - \exp(-b(T)\, t^n)] \tag{13}$$

X_{eqm} = the equilibrium amount of ferrite formed that can be formed

This equation is used to handle continuous cooling by dividing the thermal history of the steel into a number of small isothermal time segments. The fraction of the new phase formed is calculated assuming that the isothermal conditions for each time step are additive. At the beginning of each time step an equivalent time is calculated that represents the time required at the current temperature for the volume fraction present at the end of the previous time increment to form. This is done by rearranging equation (13) to give:

$$t_{eq} = -\frac{1}{b(T)} \ln\left[1 - \frac{X_{eqm}(\text{old})}{X_{eqm}(T)}\right]^{1/n} \tag{14}$$

where:

t_{eq} = the equivalent time at the current temperature for the amount of (T) phase formed at the previous time and temperature to form,

$X_{eqm}(T)$ = the equilibrium amount of ferrite that can be formed at the current conditions,

$X_{eqm}(\text{old})$ = the equilibrium amount of ferrite that could be under the previous conditions,

B, n are constants for the current temperature.

This equivalent time is then incremented to become the equivalent holding time at the new temperature, and the volume fraction transformed is calculated using equation (13).

The formation of martensite is modelled by first calculating the martensite start temperature from the chemistry of the steel, using equation (15).[26] This equation is empirical in nature and was regressed from data on 184 steels with a wide range of chemistries.

This equation has the form:

$$M_s = 512 - 453C - 16.9\,Ni + 15\,Cr - 9.5\,Mo + 217\,C^2 - 71.5\,C\,Mn - 67.6\,C\,Cr \tag{15}$$

If there is austenite that has not been transformed to ferrite and pearlite, the volume fraction of martensite formed is then calculated based on the temperature of the steel below the martensite start temperature. The equation used to calculate the volume fraction of martensite was developed by Koistinen and Marburger[31]. This equation is shown below:

$$X_M = 1 - \exp - [0.011\,(M_s - T)] \tag{16}$$

154 Quenching and Carburising

where:

X_M = the volume fraction martensite,
M_S = the martensite start temperature given in equation (15) above,
T = the current temperature.

In the generation of each new phase there can be large amounts of heat generated in the steel due to the latent heat of transformation. This is particularly true in steel when austenite is transforming to pearlite. For this reason the transformation model is closely linked to the thermal evolution model so that the thermal history of the steel and the rate of transformation can be accurately predicted. As explained in a previous section, the latent heat of transformation for any new phase is calculated at the end of each time step. This is done by calculating the latent heat of transformation at the current temperature and then applying equation (17) to obtain the heat generated per unit volume.

$$q = \rho \cdot \Delta H \cdot \Delta X / \Delta t \quad (\text{Jm}^{-3}) \tag{17}$$

where:

q = the heat generated per unit volume,
ΔH = the latent heat of transformation,
ΔX = the fraction of new phase formed,
ρ = density,
Δt = the time increment.

From the heat generated the temperature rise in the steel is calculated by using the thermal properties of the current phases present in the steel. This temperature rise is then added to the temperature of the steel. Provided that the time step is small this method gives excellent predictions of the steel thermal history and microstructure evolution.

Examples of modelled and measured cooling curves for low (0.06) and high (0.8)%C–1% Mn steels show that the evolving microstructure significantly alters the cooling curve of the steel, and that the integrated microstructural-thermal models are able to reproduce the cooling behaviour (Fig. 7). For the low carbon steel there is relatively little heat of transformation and the distinct change in cooling rate is due to the change in thermophysical properties; in particular, the specific heat. The ferromagnetic–paramagnetic transformation effects on the specific heat must also be included for this steel. For the eutectoid steel the change in cooling behaviour is almost entirely due to the heat of transformation with a smaller effect from the difference in thermophysical properties between the austenite and pearlite.

The work to date indicates that the control of the post-deformation cooling poses many problems. The strong temperature dependance of the heat transfer coefficient combined with the effects of transformation on cooling history make it almost impossible to use simple models to accurately predict the stop cooling temperatures used in modern multistage cooling. Therefore, for these new TMCP techniques there are a number of advantages in having detailed models of the physical processes occurring. These can then be used to regress more simple

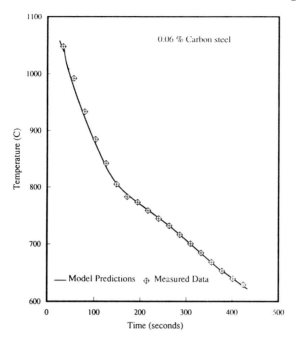

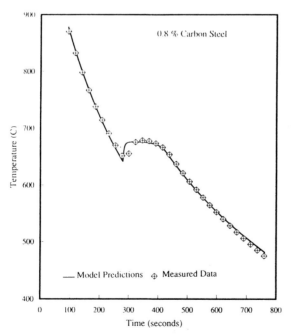

Fig. 7. Measured and predicted centre-line temperature for the air cooling of (a) 0.06 and (b) 0.8 C 20 mm plate.

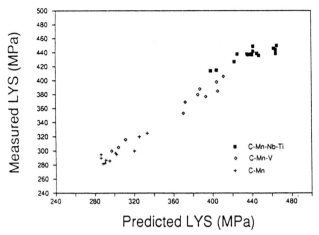

Fig. 8. Comparison of model predictions for strength of finished product with measured values.

control models or to evaluate the effects of change in composition, hardware and processing conditions.

3.2.2 Ferrite grain size prediction

In many of the steels and processes modelled to date it is the final average ferrite grain size which determines the mechanical properties. The factors that affect the ferrite grain size are: initial austenite grain size, retained strain, composition and cooling rate. The initial austenite grain size relates to the last fully recrystallised grain size, while the retained strain applies to that strain not removed by recrystallisation prior to transformation. In C-Mn, C-Mn-V and C-Mn-Ti steels there is a strong effect of composition on the ferrite grain size. The model for ferrite grain size in C-Mn steels (Table 1) is an extension of the equation proposed by Sellars and Beynon.[32] Here the effects of grain size and cooling rate are additive, while the effect of retained strain is incorporated by multiplying the ferrite grain size calculated in the absence of retained strain, $d\alpha_o$, by a function of the retained strain, ε_r,[32] (equations (9) and (10) in Table 1).

Therefore, composition is seen to affect both the base ferrite grain size and the contribution from cooling rate. It has recently been observed that as the C_{eq} increases above 0.35 the composition/cooling rate interaction disappears (i.e. $\beta_3 = 0$). Hence, different constants are used for the two composition ranges (Table 1). Both Ti (<0.02) and V (<0.06) do not appear to affect the ferrite grain size directly.

For C-Mn-Nb and C-Mn-Nb-Ti steels the ferrite grain size is found to be predominantly determined by retained strain, while the effect of composition is less important. Equation (11) in Table 1, developed for predicting the ferrite grain size after plate rolling, considers retained strain, ε_r, as an integral factor rather than a separate modifying factor, as discussed in the equation for C-Mn steels. This equation consists of a limiting ferrite grain size, $d\alpha_{lim}$, and the sum of a cooling rate term and an austenite grain size term. At strains of greater than 1.6 the effect of retained strain is reduced[10] and this term is not used to further modify the grain size.

4 STRUCTURE PROPERTY RELATIONSHIPS

For low C steels where the strength is predominantly a function of the composition and the final ferrite grain size the following equations have been obtained for the lower yield stress (LYS) and tensile strength (TS)[3,9]:

$$LYS = A_1 + B_1[Mn] + C_1[Si] + \ldots + \sigma_{ppn} + k_{1y}\, d\alpha^{-0.5} \quad (MPa) \quad (18)$$

$$TS = A_2 + B_2[C] + C_2[Mn] + \ldots + \sigma_{ppn} + k_{2y}\, d\alpha^{-0.5} \quad (MPa) \quad (19)$$

The agreement between model and measurement for a range of low C steels and hot rolled products is shown in Fig. 8.

For higher C steels the effect of pearlite is incorporated using the equations by Gladman et al.[33] At present the strength models for multiphase steels are still under development. One approach proposed elsewhere,[34] which is being investigated, is to relate the tensile properties of these steels to the average transformation temperature of that phase and to then use a law of mixtures to obtain the bulk tensile properties. A similar law of mixtures approach has been used by the current authors to predict the strength of quench and self tempered steels with some success. Here the hardness change due to tempering is modelled using a temperature-time equivalence similar to that proposed by Blondeau et al.[35]

5 CONCLUSIONS

A mathematical model has been developed to follow the thermal and microstructural evolution of steel during thermomechanical processing. While the model has been predominantly applied to hot rolling it may also be applicable to hot forging and reheat followed by controlled continuous cooling. It has been shown that the evolution of microstructure affects the temperature through changes to the thermophysical properties and the heat evolved during the formation of each phase. With this integrated approach to modelling it is possible to design new compositions and processing routes to achieve a given package of properties. It is also possible to use this model to evaluate proposed modifications to current practices and hardware.

ACKNOWLEDGEMENTS

The authors would like to thank The Broken Hill Proprietary Company Limited for permission to publish this work. They also acknowledge the support and assistance provided by many people both at these laboratories and elsewhere throughout the Company.

REFERENCES

1. R.E. Gloss, R.K. Gibbs and P.D. Hodgson, elsewhere in this conference.
2. P.D. Hodgson and R.K. Gibbs: 'Step into the 90's', *Transactions of a Joint Conference on Corrosion, Finishing and Materials*, A. Atrens et al. eds, IMMA, 1989.
3. P.D. Hodgson, J.A. Szalla and P.J. Campbell: *4th Int. Steel Rolling Conference, Deauville, France*, 1987, paper C8.
4. L.D. McKewen and P.D. Hodgson: *Proc. Materials Technology Conf., Inst Metals and Materials Australasia*, 1986, paper 3-B-9.
5. L.D. McKewen, R.K. Gibbs and B. Gore: *32nd Mechanical Working and Steel Processing Conference Proceedings, ISS-AIME*, 1991, p. 87.
6. J.A. Szalla, G. Glover and P.M. Stone: *29th Mechanical Working and Steel Processing Conference Proceedings, ISS-AIME*, 1988, p. 87.
7. C.M. Sellars: *Hot Working and Forming Processes*, C.M. Sellars and G.J. Davies, eds, Metals Society, London, 1980, p. 3.
8. C.M. Sellars and J.A. Whitemen: *Metal Science, 1979* **13**, 187–194.
9. P.D. Hodgson and R.K. Gibbs: *Int. Symp. Mathematical Modelling of Hot Rolling of Steel*, S. Yue, ed., C.I.M.M., 1990, p. 76.
10. R.K. Gibbs, P.D. Hodgson and B.A. Parker: *Morris E. Fine Symposium*, P.K. Liaw, I.R. Weertman, H.L. Marcus and J.S. Santer, eds, TMS, 1991, p73.
11. W. Roberts, A. Sandberg, T. Siwecki and T. Werlefors: *HSLA Steels: Technology and Applications*, ASM, 1984. p. 67.
12. C.I. Garcia, A.K. Lis and A.J. DeArdo: *Proc. Int. Symp. on Microalloyed Bar and Forging Steels*, M. Finn, ed. CIM, 1990, p. 25.
13. P.D. Hodgson, R.E. Gloss and G.L. Dunlop: *32nd Mechanical Working and Steel Processing Conference Proceedings, ISS-AIME*, 1991, p. 527.
14. M. Mitsutsuka: *Tetsu-to-Hagane, 1983*, **69**, 268.
15. A. Ohnishi, H. Takashima and M. Hariki: *Trans. ISIJ, 1987*, **27**, B299.
16. M. Mitsutsuka: *Tetsu-to-Hagane, 1968*, **54**, 1457.
17. H. Muller and R. Jeschar: *Arch. Eisenhuttenwes., 1973*, **44**, 589.
18. E.A. Mizikar: *Iron and Steel Engineer, June 1970*, p. 53.
19. K. Sasaki, Y. Sugatani and M. Kawasaki: *Tetsu-to-Hagane, 1979*, **65**, 90.
20. H. Kamio, K. Kunioka and S. Sugiyama: *Tetsu-to-Hagane, 1977*, **63**, S184.
21. M. Mitsutsuka and K. Fukuda: *Tetsu-to-Hagane, 1983*, **69**, 262.
22. E.B. Hawbolt, B. Chau and J.K. Brimacombe: *Metall. Trans., 1983*, **14A**, 1083.
23. E.B. Hawbolt, B. Chau and J.K. Brimacombe: *Metall. Trans., 1985*, **16A**, 565.
24. M. Umemoto, N. Nishioka and I. Tamura: *J. Heat Treating, 1981*, **2**, 130.
25. J.S. Kirkaldy and B.A. Baganis: *Metall. Trans., 1978*, **9A**, p. 495.
26. K.W. Andrews: *JISI, 1965*, **203**, p. 721.
27. E. Schiel: *Arch. Eisenhutten. 1935*, **12**, p. 565.
28. M. Avrami: *J. Chem. Physics, 1939*, **7**, p. 1103.
29. M. Avrami: *J. Chem. Physics, 1940*, **8**, p. 212.
30. M. Avrami: *J. Chem. Physics, 1941*, **9**, p. 177.
31. D.P. Koistenen and R.E. Marburger: *Acta Met., 1959*, **7**, p. 59.
32. C.M. Sellars and J.H. Beynon: 'High Strength Low Alloy Steels', *Proc. Conf. Wollongong Australia 1984*, ed D. Dunne and T. Chandra, South Coast Printers, 1985, p. 142.

33. T. Gladman, I.D. McIvor and F.B. Pickering: *JISI, 1972*, **210,** p. 916.
34. H. Yada: *Proc. Int Symp. Accelerated Cooling of Rolled Steel*, ed. G.E. Ruddle and A.F. Crawley, Pergamon Press, 1987, p. 105.
35. R. Blondeau, Ph. Maynier, J. Dollet and B. Vieillard-Baron: *Heat Treatment '76*, The Metals Society, 1976, p. 189.

10

Measurement and Characterisation of Air-Mist Nozzles for Spray Quenching Heat Transfer

M.S. JENKINS, S.R. STORY and R.H. DAVIES

BHP Research, Melbourne Laboratories, 245 Wellington Rd., Mulgrave, Victoria, Australia 3170

ABSTRACT

This paper describes a series of tests that may be used to characterise water spray quenching nozzles, with particular emphasis on air-mist nozzles. These tests include the measurement of water and air flowrate, droplet size distribution, water distribution across the spray pattern and surface heat transfer coefficients. Droplet size distributions were analysed using laser diffraction techniques. Spray water distribution was analysed using two different patternator designs. Surface heat transfer coefficients were measured by both steady-state and unsteady-state techniques. Of these two techniques for heat transfer measurement, the unsteady-state technique is time consuming, but more convenient for generating the overall heat transfer coefficient versus surface temperature relationship. The steady-state technique on the other hand is relatively quick but cannot be used below the Leidenfrost temperature. Using the results of all these tests it is possible to characterise and compare the performance of various spray nozzles. The importance of the various spray parameters can be established, and from this, decisions about nozzle selection and optimum operating conditions can be made.

1 INTRODUCTION

Quenching hot metal surfaces with vaporising liquids is used in many processes to obtain the desired material properties. The process takes advantage of the very high rates of heat transfer associated with liquid vaporisation. Quenching with liquids can be divided into essentially three basic techniques: *Immersion quenching*, where metal specimens are dipped into baths of different liquids; *film quenching*, where the cooling liquid is poured over the specimen so that a water film runs down its side; and *spray quenching*, where the metal surface is impinged with spray jets.

The main difference between spray quenching and the other techniques is that it has the potential for close control over a wide range of heat transfer rates. The controlled cooling rates offered by spray quenching are often required to obtain the desired material properties, and therefore a knowledge of the various factors affecting spray heat transfer is important. To this effect, this paper describes some techniques used to characterise spray nozzles, with particular reference to air-mist nozzles.

The degree of cooling effected by any nozzle is primarily determined by the water flux incident on the cooled surface. Other factors, however, can modify the cooling efficiency of the spray, such as droplet size and droplet velocity.[1,2] Surface temperature is also important as it governs the mechanism of heat transfer and is of particular significance below the Leidenfrost temperature.[3] In selecting a nozzle capable of satisfying the requirements of a particular application, it is often useful to have an appreciation of the effects of nozzle design parameters on nozzle performance. The nozzle test techniques summarised in this paper aim to provide the following information to help improve this understanding:

- nozzle pressure-flow relationships (referred to as performance curves);
- spray water distribution in the through-width and through-thickness direction;
- droplet size distributions;
- surface heat transfer coefficients.

2 EXPERIMENTAL

2.1 PERFORMANCE CURVES

For air-mist nozzles, nozzle performance curves are generated by measuring water and air flowrates over the full range of air and water operating pressures (referred to as Pa and Pw respectively) and an example is shown in Fig. 1. Water flowrates are measured using a turbine flowmeter and a stop-watch. Air flowrates are recorded using a precision bore rotameter. Both the air and water flowmeters are located within 1 m of the nozzle tip.

2.2 DROPLET SIZE TESTS

A knowledge of droplet size distribution as a function of the nozzle operating conditions is a pre-requisite for the fundamental analysis of heat transfer. However, despite the importance of droplet size, very little work has been carried out to determine the effect of droplet size on the heat transfer coefficients. This is due primarily to the complex nature of the problem, as well as difficulty in obtaining accurate and meaningful measurements of all the parameters involved in the heat transfer mechanism. Therefore, there is a need to investigate the relationship between droplet size and the various nozzle operating parameters.

Over the years, a wide range of techniques have been developed for determining the droplet size distribution in sprays.[4] In the current work, however, a method

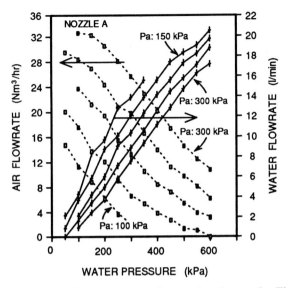

Fig. 1. An example of a performance curve for an air-mist nozzle. The dashed lines represent air flowrate while the solid lines denote water flowrate.

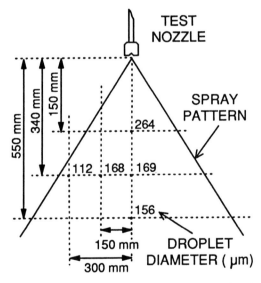

Fig. 2. Results of particle size analysis for an air-mist nozzle showing the variation in droplet size for different locations within the spray pattern.

based on light diffraction was used.[5] Droplet size measurements were obtained using a MALVERN 2200/3300 particle sizer. Figure 2 shows the results of several droplet size analyses for an air-mist nozzle at various positions within the spray

164 *Quenching and Carburising*

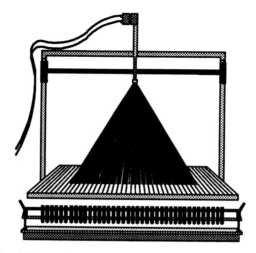

Fig. 3(a). Through-width patternator.

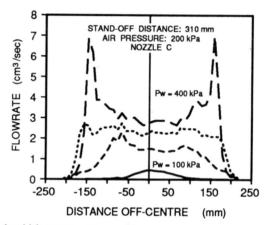

Fig. 3(b). Through-width spray water profile.

pattern. Droplets are assumed to be spherical and their size is quoted here as the median diameter based on the weight frequency distribution.

2.3 SPRAY WATER DISTRIBUTION MEASUREMENTS

The spray emitted from a spray nozzle can be designed to have almost any shaped cross-section but is usually either circular or elliptical. The shape and dimensions of the spray pattern are determined by nozzle tip design parameters and stand-off distance (SOD). The parameter which describes the amount of water reaching the surface is referred to as the water flux and has the units $l\, m^{-2} s^{-1}$. Ideally, the water flux should be uniform over the spray pattern; however, this rarely occurs in practice. Therefore, in order to obtain a meaningful assessment of nozzle performance in terms of water distribution, it is necessary to measure the quantity

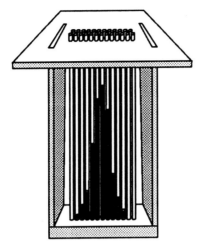

Fig. 4(a). Through-thickness patternator.

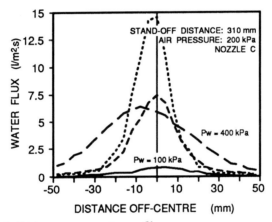

Fig. 4(b). Through-thickness spray water profile.

of water delivered by the nozzle in its characteristic directions. Circular distributions are usually symmetric about the centre of the pattern; elliptical distributions, however, exhibit two vastly different spray distributions along their major and minor axes, referred to here as the through-width and through-thickness directions respectively.

For elliptical spray patterns, measurements of spray water distribution were performed using two different 'patternator' designs. The through-width patternator consists of an inclined tray divided into a number of channels (Fig. 3(a)). Each channel is 50 mm wide and ends in a drain which directs water into a calibrated glass tube. By dividing the volume collected in each tube by the collection time, water flowrate profiles were generated. These profiles represent the average through-thickness water flux distribution of the spray pattern. An example of a through-width spray profile is shown in Fig. 3(b). Through-thickness profiles, on

166 *Quenching and Carburising*

the other hand, were generated using a smaller patternator which consists of two staggered rows of perspex tubes (8 mm diameter) covering a spray through-thickness of 100 mm (shown schematically in Fig. 4(a)). By exposing the patternator to the nozzle spray for a measured time, the through-thickness spray profile was measured. Fluxes were then obtained by dividing the volume of water collected in each tube, by both the cross-sectional area of each tube and the collection time (Fig. 4(b)).

For circular spray patterns, a patternator was used, consisting of two perpendicular rows (350 mm long) of perspex tubes (8 mm diameter).

2.4 HEAT TRANSFER TESTS

The measurement of the surface heat transfer coefficient, h, is undoubtedly the most significant factor in the assessment of nozzle performance. Much work has been done in developing techniques to investigate this quantity[3,6]; two techniques were used in this study:

- unsteady-state analysis, and
- steady-state analysis.

2.4.1 Unsteady-state heat transfer test

The unsteady-state technique is based on the measurement of sub-surface temperature-time curves in a steel block as it is spray cooled (Fig. 5(a) and (b)). The subsequent determination of h from these data involves the numerical calculation of heat flow in the block for each time interval of the logged data.[1,3,6] Analysis of the temperature data is based on the asssumption that heat flow from the sample is one-dimensional, local to the thermocouples. The analysis provides calculated values of both h and the corresponding surface temperature, at each time step.

(a) Test description. The test specimen for this technique was machined from austenitic stainless steel to a size of 120 × 120 × 70 mm, and was then instrumented with 1.5 mm diameter MIMS K-type thermocouples embedded in a staggered arrangement 2, 6 and 10 mm below the top surface. The test specimen was placed into a 15 kW muffle furnace under an argon atmosphere and soaked at 1200°C. After 40 minutes, the specimen was removed from the furnace, placed into an insulated specimen holder, which was then placed into the test position under the spray. The temperature-time profile for each thermocouple was logged by a computer controlled data acquisition system.

(b) Heat transfer model description. Surface heat transfer coefficients, based on the experimentally determined relationship between temperature at a known depth below the face of the test specimen and time (described in (a) above), are calculated using an 'inverse' solution to the transient heat conduction equation using a method of successive approximation. If the density and specific heat of the test

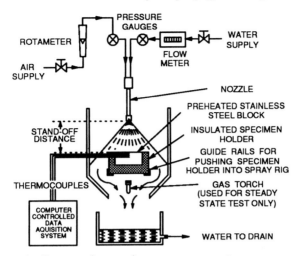

Fig. 5(a). Schematic diagram of unsteady-state test apparatus.

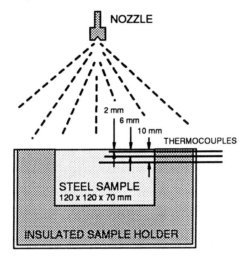

Fig. 5(b). Detail of instrumented specimen and specimen holder.

sample are assumed to be constant, and the thermal conductivity is known to vary as a function of temperature, the following 1-D heat conduction equation applies:

$$\rho C \frac{\partial T}{\partial t} = \frac{\partial}{\partial x}\left(k(T)\frac{\partial T}{\partial x}\right) \qquad (1)$$

where:
- ρ = density (kg m^{-3}),
- C = specific heat (J kg^{-1} K^{-1}),
- T = temperature (K), and
- t = time (s),
- x = distance in the x-direction (m),
- k = thermal conductivity (W m^{-1} K^{-1}).

This equation is usually solved numerically by finite difference methods (FDM) using an explicit formulation,[7,8] together with the appropriate initial and boundary conditions.

While the explicit method of solution is computationally simple, one must be careful to maintain numerical stability by the appropriate selection of Δt and Δx values. The time step, Δt, is necessarily very small because the process is valid only for certain combinations of Δt and Δx,[7,8] and Δx must be kept small in order to attain reasonable accuracy. In the current analysis, $\Delta x = 1$ mm was chosen so that the position of the second, sixth and tenth nodes of the FDM mesh were made to correspond to the 2, 6 and 10 mm thermocouple positions below the sample surface. In this way, the measured and calculated values could be directly compared. Good results were achieved by setting $\Delta x = 1$ mm and $\Delta t = 0.01$ seconds.

Between each experimentally determined pair of time and temperature values, the model assumes h to be constant. At the end of each time period, the calculated temperature was compared to the experimentally determined value. This required the use of approximate values of h, which were made progressively more accurate during successive iterations. Finally, when the difference between the calculated and experimentally determined temperatures was less than $0.5°C$, the current value of h was taken to be correct for the time period then under consideration. In the present work, the size of the time period was adjusted so that the change in temperature was always $20°C$ or more. The calculated and measured values at the 6 and 10 mm positions were used for comparative purposes only, as a check on the analysis.

(c) Analysis of cooling curve. From the 2 mm thermocouple temperature-time data, h and surface temperature were calculated at each time step using the model described above. An example of the measured temperature profiles plotted against the calculated curves is shown in Fig. 6, and an example of the model output, where the calculated h (excluding radiation) is plotted against calculated surface temperature, is given in Fig. 7. In this figure, the results of nozzle testing at three different water pressures are given.

2.4.2 Steady-state heat transfer test

The steady-state technique represents the most direct means of determining h. This technique measures h by determining the rate of heat extraction by the spray from a heated sample held at a constant surface temperature.[3,9] The method developed for the current measurements is different from the techniques previously reported (which usually employ electrical resistance heating), and is based on a system where one side of a small plate is heated by a gas flame and the heat is extracted by the spray on the other side, as shown in Fig. 8. The h is calculated from the measured thermal gradient within the plate. A constant and essentially linear thermal gradient will exist throughout the plate if constant temperatures at the thermocouples are recorded. If the positions of the thermocouples relative to the cooled surface are accurately known, the heat flux can be simply calculated.

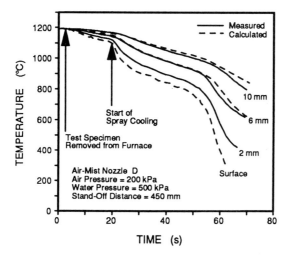

Fig. 6. Example of the calculated cooling curves superimposed over the measured curves from an unsteady-state spray heat transfer test.

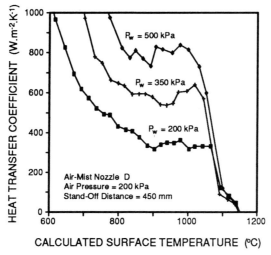

Fig. 7. Example of calculated local surface heat transfer coefficients (excluding radiation) plotted against calculated surface temperature, for three water pressures.

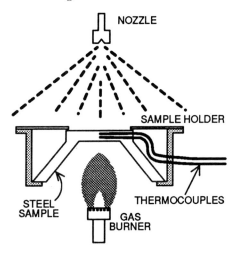

Fig. 8. Schematic diagram of steady-state apparatus.

Test description. Test specimens for this technique were machined from 8 mm thick stainless steel plate to dimensions of 8 × 45 × 55 mm. They were then instrumented with two 1.5 mm diameter, MIMS K-type thermocouples 2 and 5 mm from the top surface so that both thermocouple tips were in line and centrally located. A flame shield was also attached to the sample to protect the gas flame from overspray and to improve heating to the edges of the instrumented plate, shown schematically in Fig. 8. The instrumented test specimen was placed in the test rig, centred directly under the test nozzle and above the burner tip. When the surface of the sample reached $\approx 1000°C$ (visibly cherry red), the air and water supply pressures for the nozzle were turned on and set. The flame was adjusted such that the uppermost thermocouple temperature (T_1, 2 mm below the surface) was steady and between 950°C and 1200°C. (Below 950°C, steady-state conditions cannot be easily maintained due to the breakdown of stable film boiling at the cooled surface. Above 1200°C the sample life is dramatically shortened through scaling and/or melting.) Upon reaching steady-state (i.e. constant temperatures at both T_1 and T_2) the temperatures were recorded. The h value is calculated from the two recorded temperatures using Fourier's heat conduction equation and Newton's law of cooling (assuming 1-D heat transfer):

$$h_{tot}(T_S - T_W) = \frac{k(T) \cdot (T_2 - T_1)}{\Delta x} = \frac{Q}{A} \quad \text{(Heat flux)} \tag{2}$$

where:
 h_{tot} = total heat transfer coefficient, including radiation (W m^{-2} K^{-1}),
 $k(T)$ = thermal conductivity (W m^{-1} K^{-1}), where $T = (T_1 + T_2)/2$,
 T_S = surface temperature (K),
 T_W = water temperature (K), and
 $T_{1 \text{ and } 2}$ = thermocouple temperatures (K),
 Δx = thermocouple separation (m).

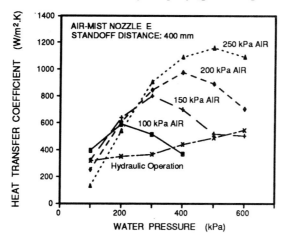

Fig. 9. Example of the results obtained from the steady-state test.

A typical set of results using this technique is presented in Fig. 9 (h values exclude radiation).

3 RESULTS AND DISCUSSION

3.1 COMPARISON OF UNSTEADY-STATE AND STEADY-STATE TECHNIQUES

A comparison of the unsteady-state and steady-state techniques is given in Table 1. It is clear from this table that the steady-state technique has several appealing benefits and is therefore the more attractive option. However, its major limitation is that it cannot be used successfully to measure h values at temperatures below the Leidenfrost temperature. In this situation, the unsteady-state technique is the preferred option. However, heat transfer rates in this region are very high, and in using the unsteady-state technique the following additional recommendations and comments are made:

- fast response thermocouples are required (faster than those used in the current work);
- data acquisition must be carried out at a high sampling frequency;
- analysis of the test results may also require a finer mesh than that suggested in Section 2.4.1.

The current specimen design used in the steady-state test (based on an 8 mm thick

172 Quenching and Carburising

Table 1. Comparison of the advantages and disadvantages of the unsteady-state and steady-state techniques for the measurement of surface heat transfer coefficients

	Unsteady-state test	Steady-state test
Advantages	More convenient method for determining h vs temperature dependence (especially below the Leidenfrost temperature).	More convenient method for determining the dependence of h on nozzle operating conditions.
	The use of three thermocouples permits a direct cross-check of the results.	3–4 minutes per test condition.
		Accuracy of the test is independent of the response time of the data acquisition system and the thermocouples.
		The specimen is not moved during the test and so no damage to the thermocouples is experienced.
Disadvantages	Inaccurate at estimating h at high heat extraction rates due to limited response time of both the thermocouples and the data acquisition system.	Careful design of the test specimen is required for estimating h at high heat extraction rates.
	Computationally difficult data analysis (numerical techniques are required).	Difficult to obtain steady-state conditions to below the Leidenfrost temperature.
	Difficulty in handling the large test specimen and the damage caused to the thermocouples by this movement.	Requires careful alignment of both spray and burner to ensure the 1-D heat flow assumption is valid.
	45–60 minutes per test condition.	Test specimens more difficult to manufacture.

plate) allows a maximum h of about 2500 W m^{-2} K^{-1} to be measured. h values above this level may cause one or more of the following problems to occur:

- melting of the sample at the hot face;
- insufficient heating capacity to achieve steady-state at surface temperatures $> 800°C$;
- excessive scaling and flaking of the specimen;
- test conditions that invalidate the assumption of one-dimensional heat flow.

The first two problems may be overcome with the use of a test specimen redesigned to incorporate a thinner plate and smaller thermocouples, but problems with the last two may still exist (this will depend mainly on test nozzle design). One method for overcoming problems with the measurement of a high local h is to test at several larger SODs (equally applicable to both techniques), and applying a non-linear extrapolation of the data to obtain h at the required SOD.

A comparison of results from the two techniques over the surface temperature range 950–1050°C were generally good, varying by only about 10%. In some cases, however, discrepancies of up to 30% have been observed. These variations have been attributed to the thickness and nature of the oxide scale on the sprayed surface of the test specimen, and problems with nozzle alignment.

3.2 THE IMPORTANCE OF AIR/WATER RATIO AND DROPLET SIZE

While water flux is the main parameter used to explain rates of heat transfer, other factors have an appreciable influence. The most significant of these secondary factors is droplet size which is related to both droplet momentum and droplet surface area to volume ratio (SA/VOL ratio). In terms of heat transfer, however, these parameters are in opposition to one another. That is, while large droplets may have an appreciable momentum, such droplets have a low SA/VOL ratio and thus the area of contact at the cooled surface will be low. Very small droplets, on the other hand, while having large SA/Vol ratios, are unable to cool effectively as they have insufficient momentum to penetrate the vapour layer at the cooled surface. Extremely fine droplets generate a 'fog' and tend to be blown away without reaching the surface.

The dilemma of simultaneously satisfying the two opposing criteria of SA/VOL ratio and momentum can at best only be solved by compromise. The literature[2] states that, in contrast to hydraulic nozzles, water droplets from air-mist nozzles are significantly smaller and achieve a higher cooling efficiency through a better compromise between SA/VOL ratio and momentum. It is often claimed that the optimum range for median droplet size is in the order of 100–200 μm.

Air-mist nozzles generate finer droplets than the conventional hydraulic nozzle due to the shearing action of the applied air in the mixing chamber of the nozzle. The literature agrees that the greatest factor affecting droplet size is the ratio of air flowrate to water flowrate (the air/water ratio, referred to here as the A/W ratio).[10] However, the relationship between A/W ratio and droplet size is not simple and is influenced by nozzle design parameters such as the air and water inlet orifice diameters, mixing chamber design and the dimensions of the nozzle tip orifice.

Tests carried out so far (typically surface temperatures above 800°C using the steady-state technique) have shown that *acceptable* air-mist nozzle operating conditions occur between an A/W ratio of 10 and 100 (below A/W=10 air-mist nozzles begin to behave hydraulically, while above A/W=100 they tend to produce a fog) corresponding to a median droplet size range of about 80–400 μm. Outside this range cooling efficiency decreases. Within this range, however, a range for optimum nozzle performance can be defined. Based on the best compromise between the high cost of compressed air and cooling efficiency, *optimum* nozzle performance has been found to occur for A/W ratios of 20–30, corresponding to a median droplet size in the range 200–300 µm.[11]

3.3. PRACTICAL APPLICATIONS

One use to which these test techniques have been applied is in the performance testing of air-mist nozzles used in the secondary cooling zones of continuous casting machines. In continuous casting, molten steel is poured into an open, water cooled copper mould where a thin shell is formed. Below the mould, air-mist nozzles are used to continue the heat extraction and solidification process. This serves to strengthen the solidified shell and to cool down the machine. Each of these functions must be achieved over a range of casting speeds without generating tensile stresses in the solid shell large enough to cause shape defects, surface cracks or internal cracks. Using the techniques described in this paper, nozzle selection procedures and operating criteria have been established. Methods for defining optimum air-mist nozzle operating conditions for use in the continuous casting process are covered in Jenkins *et al.*[11]

4 CONCLUSION

Using the results of the tests described in this paper, it is possible to determine the spray quenching performance of various hydraulic and air-mist spray nozzles. It is also possible to assess the importance of secondary factors such as droplet size on the quench cooling rate. From this knowledge more confident nozzle selections can be made, and optimum nozzle operation can be established and maintained across a wide range of cooling rates.

REFERENCES

1. E.A. Mizikar: 'Spray Cooling Investigation for Continuous Casting of Billets and Blooms', *Iron and Steel Engineer*, 1970, **47**, 53–60.
2. B. Krüger, F.P. Pleschiutschnigg and D. Zebrowski: 'The ABC's of Air-Mist Cooling', *Metal Producing*, January 1984, 37–39.
3. R. Jeschar, U. Reiners and R. Scholz: 'Heat Transfer During Water and Water-Air Spray Cooling in the Secondary Cooling Zone of Continuous Casting Plants', *Proc. 69th Steelmaking Conf., ISS-AIME*, Washington DC, 1986, 410–415.
4. A. Williams: *Combustion of Liquid Fuel Sprays*, Butterworths, 1990, 35–52.
5. P. Meyer and N. Chigier: 'Drop-size Measurement Using a Malvern 2200 Particle Sizer', *Atomisation and Spray Technology*, 1986, **2**, 261–298.
6. R.F. Price and A.J. Fletcher: 'Determination of Surface Heat Transfer Coefficients During Quenching of Steel Plates', *Met. Technol.*, 1980, **7**, 203–211.
7. G.D. Smith: *Numerical Solution of Partial Differential Equations: Finite Difference Methods*, Oxford University Press, 1985.
8. H.P. Holman: *Heat Transfer*, 6th edn, McGraw-Hill, 1986.

9. V.A. Goryainov: 'Experimental Research into Hydrodynamics and Heat Transfer During the Spray Cooling of Continuous Cast Steel Billets, Report I', *Izv. VUZ Chernaya Metall.*, *1980*, (7), 118–122.
10. L.D. Wigg: 'Drop-size Prediction for Twin-fluid Atomizers', *J. Inst. Fuel, November, 1964*, **37**, 500–505.
11. M.S. Jenkins, S.R. Story and R.H. Davies: 'Defining Air-Mist Nozzle Operating Conditions for Optimum Spray Cooling Performance', *Proc. 19th Australian Chemical Engineering Conf.*, Newcastle, Australia, September 1991, 1062–1071.

11

Investigation of Quenching Conditions and Heat Transfer in the Laboratory and in Industry

S. SEGERBERG AND J. BODIN

IVF – The Swedish Institute of Production Engineering Research, Mölndalsvägen 85, S-412 85 Gothenburg, Sweden

ABSTRACT

Programs for analysis and simulation of the hardening process have been developed over the past decade, and are constantly being developed further. A prerequisite for calculating the temperature distribution through a part is that the heat transfer coefficient between the part and the quenchant is known. This paper describes how the heat transfer coefficient varies for cylinders of differing sizes and for a disc. In addition, the cooling characteristic differs from one point of a batch to another as a result of the pattern of the batch and density of packing. This is described with examples from three different batches.

1 INTRODUCTION

Within certain limits, the characteristics of a metallic material can be modified by heat treatment. With a suitable choice of method of heat treatment, correctly carried out, it is possible to effect a significant improvement in the strength or wear resistance of a part, or to use a lower-alloyed, cheaper material.

A heat treatment operation consists of a heating and a cooling process. Both are of importance for the resulting characteristics, although generally it is the cooling sequence that is the more critical, as it must be carried out at a controlled cooling rate. A higher cooling rate generally results in higher strength and hardness, but at the same time is associated with a risk of cracking or distortion as the cooling increases. The rate of cooling must therefore be suited to the type of material and to the shape and dimensions of the product.

2 COOLING CHARACTERISTIC

The cooling sequence in hardening normally consists of three phases, as shown in Fig. 1,

178 *Quenching and Carburising*

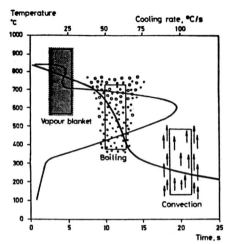

Fig. 1. Typical cooling curve for a quenchant with its three phases.

- vapour film phase
- boiling phase
- convection phase.

Figure 1 is a schematic representation of the sequence of the various phases during cooling. Heat transfer in each of the phases is by three completely different mechanisms. In the *vapour film phase*, the temperature of the surface of the metal is so high that the quenchant is vaporised and a thin film of vapour forms around the entire part. The *boiling phase* starts when the surface temperature of the part has fallen so much that radiant heat is no longer sufficient to maintain a stable vapour film. The quenchant that comes into contact with the hot surface boils instantly, removing heat at a high rate. When the surface temperature has fallen to the boiling point of the quenchant, the *convection phase* starts.

The transition from the vapour film phase to the boiling phase occurs through collapse of the vapour film at points where the temperature is lowest, e.g. at thin sections, corners and edges. In the case of a cylinder, the boiling phases starts approximately simultaneously at the corner at the two end surfaces, and then extends towards the centre, as shown in Fig. 2.

The increase in temperature of the quenchant closest to the surface of the part gives rise to convection currents carrying quenchant up the surface of the part, resulting in the two boiling fronts meeting immediately above the centre.

3 DETERMINATION OF HEAT TRANSFER COEFFICIENT

3.1 TEST PROCEDURE

The heat transfer coefficient has been determined outgoing from measurement of the cooling curve for cylindrical test probes fitted with thermocouples in the centre and close to the surface. Table 1 shows the material and dimensions of the test probes used and the positions of the thermocouples.

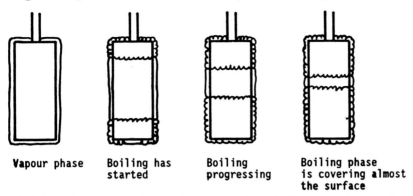

| Vapour phase | Boiling has started | Boiling progressing | Boiling phase is covering almost the surface |

Fig. 2. Schematic representation of the spread of the boiling phase in a stationary quenchant.

Cooling was done from 850°C for the Inconel test probes and from 800°C for the silver test probe. The quenchant was Isomax 166, a fast-quenching oil from Houghton. The oil temperature was 70°C and the flow rate was 0.3 m/s.

The cooling curves were measured using the 'ivf quenchotest' portable measuring unit, as shown in Fig. 3. During the experiment the standard test probe of Inconel 600 was replaced with the test probes listed in Table 1.

3.2. CALCULATION PROCEDURE

The ANSYS PC-Thermal finite element program was used to calculate the heat transfer coefficient.

Calculations were performed in stages using several different time intervals. During the first stage, the heat transfer coefficient was assumed and the temperature was calculated at the point where the thermocouple was positioned. After comparison of the result with the as-measured temperature, the heat transfer coefficient was adjusted and the calculation repeated until sufficient accuracy was obtained. This procedure was then repeated for each time interval until the heat transfer coefficient had been calculated for the entire cooling period. The lengths of the time intervals were varied: during the boiling phase, they were 0.1 s, increasing to 2.0 s at the end of the convection phase. The calculations assumed that heat conduction operated only radially, i.e. calculation was one-dimensional.

Table 1. Test probes used in the trials

Material	Dimensions, mm dia. × length	Position of thermocouple
Inconel 600	ϕ 12.5 × 60	Centre
Inconel 600	ϕ 30.0 × 90	2.0 mm below surface at half the length
Silver	ϕ 16.0 × 48	Centre

180 *Quenching and Carburising*

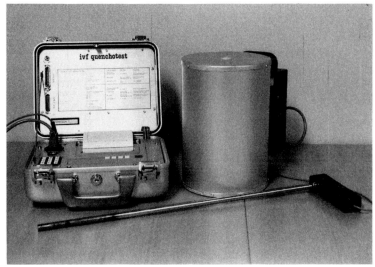

Fig. 3. The 'ivf quenchotest', as used for the experiment.

3.3 CALCULATED HEAT TRANSFER COEFFICIENT FOR CYLINDRICAL TEST PROBES

The results of the calculations for the three test probes are shown in Fig. 4.

The similarities of the heat transfer coefficients are very marked for the two Inconel test probes. The minor differences between them are fully within the calculation inaccuracy. This means that the diameters of the test probes are not critical for calculations of heat transfer coefficient within this range of dimensions, which in turn means that a heat transfer coefficient determined for one size can be used for calculation of the cooling curves of another size object made of the same material.

Agreement between the results for the Inconel and the silver test probes are less good, although the maximum heat transfer coefficients are approximately the same. The differences were mainly around 600–500°C, but also around 350–250°C, which, in the case of steel are critical temperatures for phase transformations, are so large that great care must be taken when deciding on the suitability of a quenchant for hardening steel if the performance of the quenchant has been determined using a silver test probe. The reason for this difference is probably due to different surface reactions, e.g. the emission factor, between the surface of the silver and the oil.

4 VARIATION IN HEAT TRANSFER COEFFICIENT AROUND A PART

The heat transfer coefficients, determined as described above, are valid only for the centre of the testpiece, i.e. halfway along its length. The heat transfer coefficient

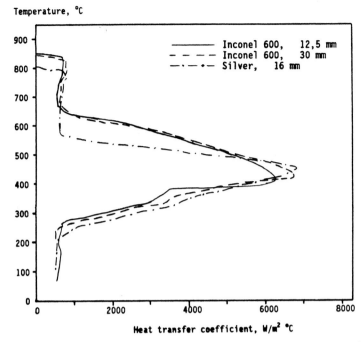

Fig. 4. Calculated heat transfer coefficients for the test probes in Table 1, based on measured cooling curves in Isomax 166 oil, 70°C and 0.3 m/s.

varies along the surface of a part. Even in the case of such a simple part as a cylinder, the heat transfer coefficient varies along its length. Figure 2 shows how the boiling phase spreads along the surface of the cylinder after the vapour film collapses first at the corners.

4.1. TEST PROCEDURE

In order to determine more accurately how the heat transfer coefficient varies from one part of a cylinder to another, trials were performed with test probes carrying four thermocouples close to the surface, as shown in Fig. 5. The figure also shows the cooling curves for each thermocouple when cooling the test probe in oil. It can be seen that the vapour film phase is longest for the thermocouple in the middle of the cylinder's length and for the corner the decrease of temperature in the vapour phase is faster due to two-dimensional heat conduction.

A Toshiba 3200 portable computer was used for temperature recording, fitted with a special card developed by IVF for simultaneous recording of temperatures from up to 16 thermocouples.

4.2 CALCULATION PROCEDURE

A newly developed program, 'ivf quenchosoft', based on the finite difference

182 Quenching and Carburising

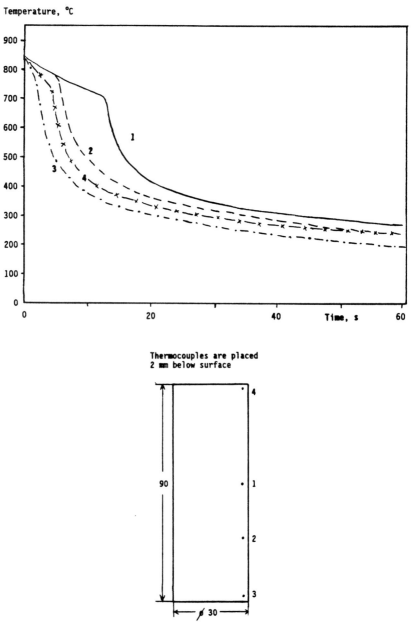

Fig. 5. Inconel 600 test probe fitted with four thermocouples, and measured cooling curves in Isomax 166 oil, 70°C and 0.3 m/s.

method, was used for the calculations. The program runs on a PC, and calculates the heat transfer coefficients fully automatically. This considerably reduces the time required for calculation in comparison with the method described in Section 3.2.

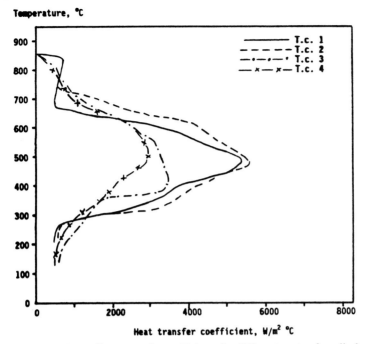

Fig. 6. Calculated values of heat transfer coefficients for different parts of a cylinder, based on data from the thermocouples positioned as shown in Figure 5.

The calculations for thermocouples 1 and 2 were made using a one-dimensional axially symmetrical model. A two-dimensional axially symmetrical model was used for thermocouples 3 and 4 in the corners.

4.3 CALCULATED HEAT TRANSFER COEFFICIENT AROUND A PART

The results of the calculations show that the heat transfer coefficients vary significantly between the corners and the side of the cylinder, as shown in Fig. 6. As expected, the vapour film phase lasts longest for the thermocouple in position 1, while the cooling performance in general for this test position is essentially the same as that for test position 2. The fact that the heat transfer coefficient for the corners is lower is probably due to the flow eddies in the oil that occur around the corners. That the upper corner has a lower heat transfer coefficient is probably due to the fact that this corner is cooled by warmer oil, i.e. oil convecting up along the cylinder and becoming hotter as it does so.

By calculating and plotting the heat transfer coefficient in this manner at points on the surface of the cylinder, it becomes possible to calculate the temperature/time curve during cooling with good accuracy for the cylinder. Knowing the temperature/time curve it then becomes possible to use more advanced calculation programs to calculate hardness, stresses and distortion. Programs of this type have been developed in Sweden and in other countries.

184 Quenching and Carburising

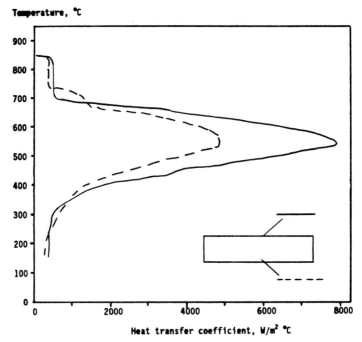

Fig. 7. Calculated heat transfer coefficient for the centre of the top and underside of a disc of Inconel 600, ϕ 90 × 20 mm. Cooling in Isomax 166 oil, 70°C and 0.3 m/s.

Another test probe for which the heat transfer coefficient has been calculated is that of a disc. The heat transfer coefficient at the centre of the top and underside are shown in Fig. 7. The fact that the heat transfer coefficient is lower on the underside is probably due to the stagnation of the oil flow that occurs beneath the disc, in combination with the effect of the vapour film/vapour bubble formed there, reducing the rate of heat transfer.

5 THE EFFECT OF THE BATCH ON COOLING

To be able to determine the heat transfer coefficient for separately cooled parts, as above, has only limited practical value. All industrial heat treating companies know that the way in which the batch is arranged, and its packing density, can have a very considerable effect on the hardening performance. This can be confirmed by measuring the hardness of test probes or items from different positions in a batch after hardening. Another, more direct, way is to measure the temperature/time curves of test probes placed at different positions in the batch. This has been done for a number of different types of batches.

5.1 TEST PROCEDURE

Test probes of Inconel 600, with dimensions ϕ 12.5 × 60 mm, were used for these trials. The thermocouples were fitted in the centre of the test probes, and with a

Table 2. Batch details

Batch no.	Parts	Quantity	Total weight (kg)	Packing density
1	Rings	1000	600	Close
2	Rings	500	350	Normal
3	Holders	25	600	Very open

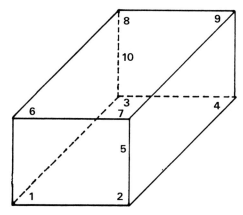

Fig. 8. Positions of the test probes in the batches. Test probes 1 to 5 in the lower layer and 6 to 10 in the upper layer.

length of about 5 m. This length was necessary to allow the test probes to follow the batch through the furnace, which was a sealed-quench type furnace with a maximum capacity of 1000 kg.

Two types of details were used: a ring with an outer diameter of 70 mm, an inner diameter of 50 mm and a length of 30 mm, and a holder with approximate dimensions of $150 \times 100 \times 250$ mm. The test probes were loaded into all the batches in an evenly distributed manner in each layer and vertically. Details of three of the batches are shown in Table 2.

The test probes, of which there were ten in each batch, were placed close to the bottom and close to the top of the batches, as shown in Fig. 8. They were positioned so as not to be in contact with the details, in order to avoid, as much as possible, being affected by the details during cooling.

The same system as described in Section 4 was used for data acquisition. Temperatures were monitored also during heating, which enabled the temperature uniformity of the furnace to be determined. After the normal holding time at the required temperature, the batch was cooled by first being transferred from the hot zone to the quench elevator above the quench tank in the vestibule and then being lowered into the tank. The hardening oil was Bellini FS at a temperature of 70°C. Agitation was the maximum possible, i.e. the agitator was running at high speed. The same agitation rate was used for all three batches.

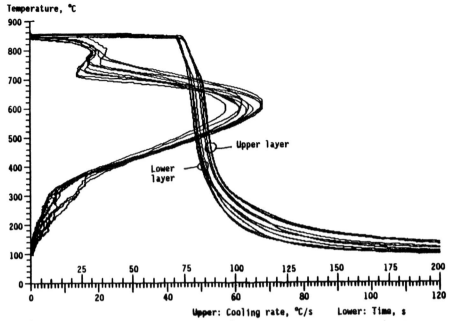

Fig. 9. Temperature/time and cooling rate curves for the test probes in Batch 1, the most closed-packed batch.

5.2 MEASURED TEMPERATURE/TIME CURVES IN THE BATCHES

Figure 9 shows the temperature/time curves and cooling rate curves at the various measurement positions in Batch 1. Temperature recording started when the batch was transferred from the hot zone to the vestibule above the quench tank. This takes just over 40 seconds and the temperature is almost constant during this time. Cooling of the lower layer of details in the batch starts a few seconds before the upper layer, as it enters the oil first. The greatest temperature difference between the test probes in the batch is somewhat over 100°C after 20–30 seconds' cooling. Even after more than one minute's cooling, this difference is still as large as about 60°C. For some details, this difference would be sufficient to give rise to variations in hardness from one position to the other within the batch. The length of the vapour phase may also be too long at some positions in the batch and so causes transformation to ferrite and perlite.

The most significant difference between the upper and the lower layers occurs during transition through the 400–200°C temperature range, which was when the test probes in the lower layer also had the highest cooling rate. The reason for the difference is of course due to the fact that, as the oil rises through the batch, it is heated and is therefore less effective in cooling the parts at the upper layer. In this respect, the packing density is very important, although any channels up through the batch also affect the efficacy of cooling. A more detailed comparison between the upper and lower layers can be made by comparing the results from test probes placed in the same corners of the batch, as shown in Fig. 10.

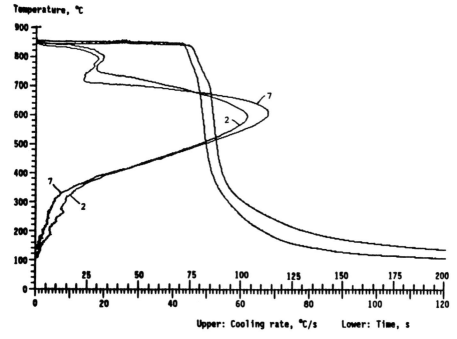

Fig. 10. Cooling rate of test probes nos. 2 and 7 (as shown in Figure 8) in Batch 1.

The difference in the cooling process between these test probes reflects the difference in cooling between the upper and lower layers in the batch. There can, of course, be departures from this pattern, but in general terms this reflects the practical reality of industrial hardening. The maximum cooling rate might vary between the upper and lower layer, but the lower layer has normally a higher cooling rate.

As previously mentioned, the packing density of the batch has a considerable effect on the cooling characteristic in different positions of the batch. A comparison of the cooling characteristic of test probe no. 10 (in the centre of the upper layer) in the three different batches (as shown in Table 2) shows a significant difference in the cooling rate as shown in Fig. 11.

Batch 1, which was the closed-packed, has the lowest cooling rate over the entire temperature range at this measurement position. Batch 2, containing half the number of rings, has a cooling rate that is about 5–15°C/s higher, and has about 30°C higher temperature than Batch 1 for a long time in the convection phase. Batch 3, containing the holders, which were very open-packed, has approximately the same maximum cooling rate as the other two batches. However, its cooling rate during the convection phase (350–100°C) is considerably higher, as the oil flow through the batch is not significantly obstructed.

6 SUMMARY

Modern computerised instrumentation equipment makes it possible to monitor

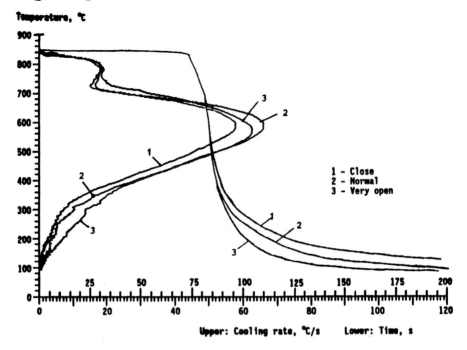

Fig. 11. The effect of the batch on the cooling characteristic in the centre of the upper layer (test probe no. 10) for the three batches.

the cooling process in much more detail and more accurately than was previously possible. At the same time, computer programs have been developed which, when run on today's powerful computers (including personal computers), permit sophisticated calculations of temperature curves and gradients, as well as phase structures, hardness and distortion after hardening. However, at the same time, it is important to remember that reliable calculations of hardening results require good quality input data and material data (coefficient of thermal conductivity, phase transformation and structure growth kinetics etc.), with measurement being made under the actual conditions to be investigated. This paper is intended to contribute to the knowledge required in connection with investigation under such actual conditions.

It is clear that not even the most advanced computer programs can produce results of greater practical value for industrial heat treatment practitioners without a better understanding of what happens around the details when they are being quenched.

ACKNOWLEDGEMENT

This work has been sponsored by the Swedish Board for Technical Development (STU), the Nordic Fund for Industrial Development (NI) and five Swedish industrial companies, to which the authors express their gratitude.

12

The Design and Performance of a Laboratory Spray Cooling Unit to Simulate In-line Heat Treatment of Steel

R.E. GLOSS, R.K. GIBBS and P.D. HODGSON

BHP Research, Melbourne Laboratories, Mulgrave, Australia 3170

ABSTRACT

The design and performance of a laboratory spray cooling system is discussed. The facility has been used in conjunction with an experimental rolling mill to simulate a number of thermomechanical controlled processes applied to steel products, including recrystallisation controlled forging followed by direct quenching and self tempering and 3 stage cooling of steel strip. The ability of the system to produce a broad range of cooling rates from 9 to 50°C/s while cooling 12 mm plate has also been shown. The development of a mathematical model which predicts the temperature of the plate during rolling and spray cooling was also undertaken for this facility. Excellent agreement between the temperature model and measured data for a range of conditions was achieved.

1 INTRODUCTION

Thermomechanical controlled processing (TMCP) or thermomechanical treatment (TMT) is now being widely applied to many manufacturing technologies. TMT is based on the control of the microstructure during deformation and cooling of the workpiece following deformation; preferably with little or no reheating. Applications range from recrystallisation controlled forging (RCF) followed by direct quenching and self tempering (QST) to the complex multi-stage run-out table cooling required to produce modern high strength, high elongation hot rolled strip steels.

In order to investigate and develop TMT strategies, a laboratory spray cooling system has been developed to simulate the heat treatment of a wide range of products ranging from rapid quenching and self tempering, typical of double quench forged products, to the more moderate cooling rates used in accelerated cooling of thick, hot rolled plate.

190 *Quenching and Carburising*

Fig. 1. Spray cooling unit.

To realise this range of cooling rates, water sprays capable of providing a wide range of heat transfer conditions are required. The sprays chosen use a mixture of air and water at varying pressures to alter water droplet size, velocity and flux to the workpiece surface and, hence, the heat transfer coefficient. Most of the testing was performed on flat plates of the steel under investigation, which were hot rolled on the adjacent laboratory mill. Different plate thicknesses were used to simulate the variation in section thickness found in more complex product geometries.

2 RIG DESIGN

The cooling unit is located adjacent to the two high reversing laboratory mill. Figure 1 shows a general view of the cooling unit. The major components are:
- the specimen cooling system,
- the specimen transfer system,
- the temperature measurement system.

The cooling unit is 6 m in length, of which the centre 3 m are cooling zones. Due to space constraints the unit is attached at 90 degrees to the existing run-out table of

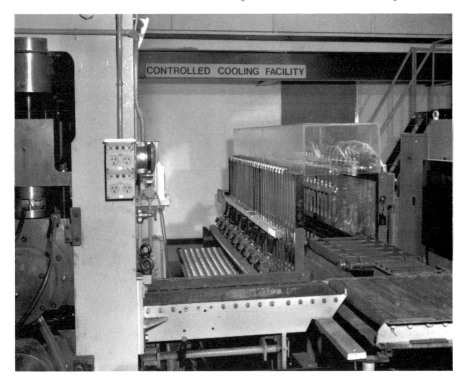

Fig. 2. Position of the cooling facility.

the Marshall-Richards rolling mill (Fig. 2). While not ideal, this position still allows fairly rapid transfer of most hot rolled specimens to the cooling rig after rolling. For critical investigations where the time between rolling and cooling must be minimised to a few seconds, it is possible to move the entire unit in-line with the rolling mill.

2.1. COOLING SYSTEM

This is the main section of the unit and has been designed to give maximum flexibility in achieving a wide range of cooling rates for a given thickness. It consists of nine cooling bank modules, one of which is shown in Fig. 3. Each assembly consists of 6 nozzles (3 top, 3 bottom), with flow and pressure adjustment for both water and air. For 36 of the 54 spray nozzles a 3/2 solenoid valve has been connected in series with the spray nozzles to allow dynamic closing of the nozzle and redirecting of the water flow through the third port in the valve. This configuration allows the nozzles to be switched in or out while maintaining flow and pressure equilibrium in the system. The individual assemblies can be moved along the supporting frame to give some variation in distance between banks. The nozzles with the 3/2 valves can be placed anywhere within the 3 m cooling zone.

192 Quenching and Carburising

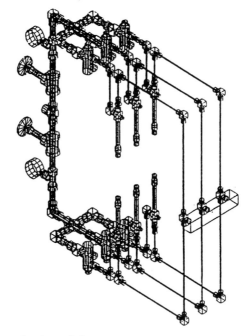

Fig. 3. Detail of spray bank module.

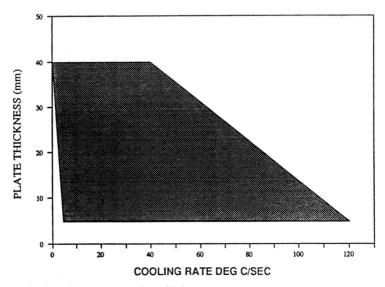

Fig. 4. Desired cooling rates vs. plate thickness

The cooling facility has been designed to cover the range of the continuous cooling rates for the thicknesses shown in Fig. 4 which encompasses those required for a range of products from QST sections to soft accelerated cooled plate. It has also been designed to be capable of simulating complex cooling strategies. For example, multi-stage cooling patterns for microstructure control in hot strip rolling, and double quench processes for recrystallisation controlled forging operations.

2.2 SPRAY NOZZLES

The spray nozzle used in this facility is the Casterjet 3/8 CJL-5-120. A detailed characterisation and performance study was undertaken on this nozzle[1] prior to its choice. This worked showed this nozzle to be suitable as:

- it has a large turn down ratio while still providing a uniform water flux distribution (Fig. 5);
- it has the versatility of being used as air only (for thin sections) or as an atomising spray which can be adjusted for a wide range of cooling rates; and
- it provides effective heat transfer.

In total the system has 54 nozzles requiring, at the top of their performance (i.e. 8 bar water, 4 bar air), the delivery of 1200 l/min of water and 28 m^3/min air. The water is delivered from a centrifugal pump coupled to a catchment tank providing a fully recirculating system. The high volume air requirement is delivered from a 25 m^3/min compressor.

2.3 SPECIMEN TRANSFER SYSTEM

The unit has been designed to carry plate lengths up to 2 m and thicknesses from 5 mm up 40 mm with a maximum width of 200 mm. The sample is placed on a trolley which has centreing guides to ensure that it is correctly aligned with the spray nozzles. It is supported on a series of removable thin plates which ensure minimum interference with the bottom set of nozzles. The trolley is driven by a cable-pulley arrangement, coupled to an AC reversing variable speed motor gearbox assembly. This allows controlled pull-through speeds from 0.01 m/s to 1 m/s.

2.4 TEMPERATURE MEASUREMENT

The temperature of the steel sample during rolling and through the cooling zone can be measured by two type N thermocouples embedded in the plate before it is reheated. These are attached to retractable reels preventing tangling of the thermocouple compensating leads during the trolley traverse. In addition to the thermocouples an Australian Optical Fibre Research – Miniature Dual Wavelength pyrometer (AOFR MDW 1000) can be used for surface temperature measurement in the cooling zone. The range of this unit is 500–1500°C with a

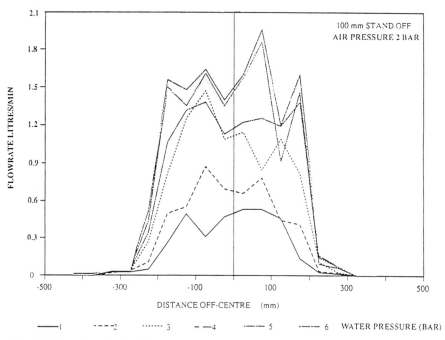

Fig. 5. Water flux distribution.

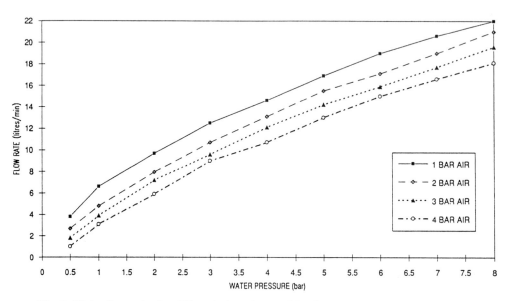

Fig. 6. Water flow rates for different air water combinations.

resolution of 1°C and response time of less than 1 s. The unit has four separate probes which are placed on the unit at 1 m intervals, so that the surface temperature of the plate can be measured at the entry and exit positions and two positions in the spray unit.

3 COOLING CHARACTERISATION

3.1 NOZZLE PERFORMANCE

Figure 6 shows the water flow rates for water pressure ranging from 0.5 bar to 8 bar and for 1, 2, 3, and 4 bar air pressures, respectively. From this it can be seen that water flow rates from as high as 22 l/min per nozzle down to 1 l/min per nozzle are possible. The integrity of the spray nozzle pattern was maintained over these pressure and flow rate combinations.

3.1 CONTINUOUS COOLING

A series of tests was undertaken using low C steel plates 24 mm thick × 100 mm wide and 100 mm in length with two 1.5 mm thermocouples embedded at mid-thickness at the centre of the plate. The plates were soaked at 1230°C for 1 hour and then rolled according to Schedule 1 outlined in Table 1, and cooled using the cooling rig set-up conditions in Table 2. Figure 7 shows the resultant temperature–time traces for all conditions. For a constant air pressure of 2.0 bar it is possible to maintain the integrity of the spray for water pressure from 0.5 to 8.0 bar which gives cooling rates from 10°C/s to 40°C/s for the 12 mm plates tested. The complete results for the cooling rates measured between the critical range of 850–600°C are shown in Table 3.

The effect of air pressure was investigated for setups 6 and 7 (Table 2). Increasing the air pressure from 2 to 4 bar increased the cooling rate from approximately

Table 1. Rolling schedules

Pass No.	Roll gap (mm)	Reduction (%)	Exit temperature (°C)
Schedule No. 1			
Start	24	–	1230
1	17	29.2	1150
2	12	29.4	1080
Schedule No. 2			
Start	27		1250
1	15	44	1050
2	5	67	950

Table 2. Spray right setups

	Cooling rig water/air pressure setup conditions (bar)																	
	Zone 1						Zone 2						Zone 3					
	Bank 1		Bank 2		Bank 3		Bank 4		Bank 5		Bank 6		Bank 7		Bank 8		Bank 9	
Sample ID	W	A	W	A	W	A	W	A	W	A	W	A	W	A	W	A	W	A
1	0.5	2	0.5	2	0.5	2	0.5	2	0.5	2	0.5	2	0.5	2	0.5	2	0.5	2
2	1	2	1	2	1	2	1	2	1	2	1	2	1	2	1	2	1	2
3	2	2	2	2	2	2	2	2	2	2	2	2	2	2	2	2	2	2
4	4	2	4	2	4	2	4	2	4	2	4	2	4	2	4	2	4	2
5	8	2	8	2	8	2	8	2	8	2	8	2	8	2	8	2	8	2
6	2	4	2	4	2	4	2	4	2	4	2	4	2	4	2	4	2	4
7	8	4	8	4	8	4	8	4	8	4	8	4	8	4	8	4	8	4

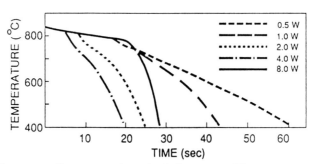

Fig. 7. Continuous cooling curves at constant air pressure 2 bar.

Table 3. Continuous cooling results

	Setup		Cooling rate 850–600°C
Sample	Air (bar)	Water (bar)	(°C/sec)
1	2	0.5	9
2	2	1	15
3	2	2	22
4	2	4	28
5	2	8	40
6	4	2	29
7	4	8	56

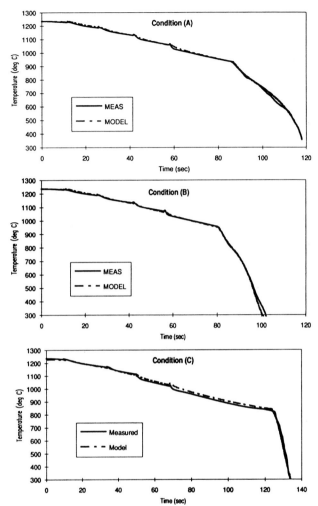

Fig. 8. Agreement between model predictions and measured data for conditions (A) 0.5 bar W, 2.5 bar A, (B) 2 bar W, 2 bar A and (C) 8 bar W, 4 bar A.

20°C/s to 30°C/s and from 40°C/s to over 50°C/s for the water pressures of 2.0 and 8.0 bars, respectively. Therefore, it can be seen that air pressure does have an appreciable effect on cooling rate, and that it can be kept constant at 2.0 bar for most tests. However, where low cooling rates ($<10°C/s$) are required for thin sections (<5 mm), low water pressures (<0.5 bar) must be used and the air pressure must be increased to maintain integrity of the sprays.

3.3 MODELLING

During the characterisation of the cooling unit the large amount of data collected were used to develop and calibrate a mathematical model which predicts the

198 Quenching and Carburising

temperature of the plate during rolling and spray cooling. The temperature model for rolling was adapted from a one dimensional implicit finite difference model originally developed for a full scale plate mill.[2] No significant changes were required to accurately predict the temperature during the rolling and air cooling phases of the process in the laboratory mill. However, as is to be expected, the heat transfer coefficient used during the spray cooling section required further development.

It was not possible to use a constant heat transfer coefficient throughout the spray cooling section because the temperature range covered by the rig is so large. The heat transfer coefficient for purely hydraulic water spray has been determined separately and the results are presented elsewhere in this conference.[3] The heat transfer coefficient is calculated as function of temperature and water flux using equation (1). It is this equation which forms the basis of the heat transfer calculations in the spray cooling section of the work.

$$h_c = A \times \dot{w}^B [T_s - n(T_s - T_n)]^c \times m \tag{1}$$

where

$$m = 1 - \frac{1}{\exp\{(T_s - T_m)/M\} + 1}$$

$$n = 1 - \frac{1}{\exp\{(T_s - T_n)/N\} + 1}$$

and

$\dot{w}$ = water flux rate (1 m^{-2} s),
T_s = surface temperature (K),
T_m = temperature of maximum heat transfer (K),
T_n = temperature above which h_c becomes constant (K),
A, B, C, M, N = constants.

The water flux rate for the current nozzles is calculated using equations (2) and (3) which are derived from an empirical fit for the data shown in Fig. 6.

$$\dot{w} = \frac{WF}{\text{Area}} \tag{2}$$

$$WF = 5.5 \left[\frac{W^{0.65}}{A^{0.15}} \right] \tag{3}$$

where:

WF = water flow (l/min),
A = air pressure (bar),
W = water pressure (bar),
Area = area of the spray at the workpiece surface (m^2).

As equation (1) is strictly valid for hydraulic nozzles and the nozzles used in the cooling rig are atomising air/water nozzles it was expected that some modification to this equation would be required. It is known that the heat transfer efficiency obtained from atomising nozzles is higher than purely hydraulic nozzles and for these nozzles the manufacturer claims that they are about twice as effective.[4] However, the best fit between the measured data and the predicted cooling curves (Fig. 8) was achieved with a heat transfer coefficient three times that calculated by equation (1).

An accurate temperature model of the rolling and cooling is a powerful tool which can be used to reduce the number of experiments required to achieve a given cooling scenario. This is particularly important for multistage cooling processes where the cooling stop temperatures are as important as the cooling rates. The model also allows the accurate planning of rolling schedules to achieve the correct controlled deformation temperatures and cooling start temperatures for the process being simulated.

The temperature model is also an integral part of the process controller and data logging system which has been designed to enable research into feed-forward and feed-back online control systems for cooling processes. This will allow computer control of the switching of the nozzle setup to ensure that the required cooling start temperature and cooling stop temperature are obtained.

4 APPLICATIONS

To demonstrate the flexibility and effectiveness of the unit it was decided to simulate a QST process and a three-stage cooling process. These two simulations were chosen because they require tight control over cooling rates and cooling start and finish times and, in the case of QST, very fast cooling rates.

4.1 QST PROCESSING

The QST process requires extremely rapid cooling of the surface layers, whilst retaining sufficient heat in the centre of the workpiece to allow the recalescence to reheat and thus temper the surface layers. If the cooling on the surface is not sufficiently severe, too much heat will be extracted from the centre section as the surface is cooled to the required temperature.

Several trials were carried out using a 100 mm long by 80 mm wide by 40 mm thick samples rolled to 16 mm and passed through the cooling rig. In the initial tests the cooling rig was operated with the condition shown in Table 4 (Sample QST1) with a trolley speed of 1 m/sec. Using these conditions it was possible to obtain the desired cooling rate at the centre followed by self tempering of the surface layers due to residual heat in the core. However, inspection of the microstructure showed that typical QST structure had not been obtained.

When the cooling rig is operated in normal setup there are appreciable gaps between the areas where the water sprays impinge on the surface of the plate. This causes small oscillations in the cooling rate as the plate traverses through the sprays. For general continuous cooling work these oscillations are minor and can be neglected as it is the average cooling rate from 850°C to 600°C that is important in determining the microstructure of the product. However, in the extreme case of

200 *Quenching and Carburising*

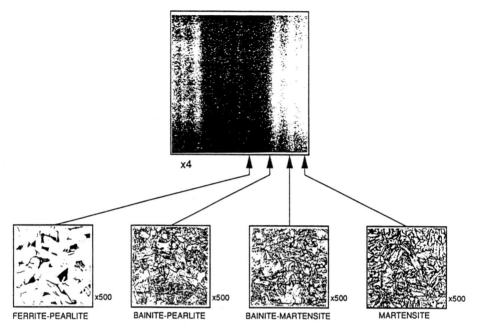

Fig. 9. Microstructure of QST section produced on laboratory cooling unit.

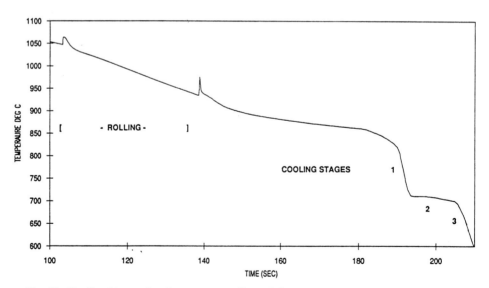

Fig. 10. Cooling history for three-stage cooling trial.

Table 4. Setup for QST and three-stage cooling

	Cooling rig water/air pressure setup conditions (bar)																	
	Zone 1						Zone 2						Zone 3					
	Bank 1		Bank 2		Bank 3		Bank 4		Bank 5		Bank 6		Bank 7		Bank 8		Bank 9	
Sample	W	A	W	A	W	A	W	A	W	A	W	A	W	A	W	A	W	A
QST1	0	0	10	2	10	2	10	2	10	2	10	2	10	2	10	2	0	0
QST2	10	2	10	2	10	2	0	0	0	0	0	0	0	0	0	0	0	0
Three-stage cooling	0.4	3	0.4	3	0.4	3	0	0	0	0	0	0	0.4	3	0.4	3	0.4	3

QST these small oscillations become important because they allow the surface to be reheated between sprays, which, in turn, slows the overall cooling rate at the surface to such an extent that the thin layer of martensite (in the case of this steel) cannot form and the resultant microstructure is made up of softer transformation products (e.g. bainite).

To achieve the required cooling rates, three banks of nozzles were angled together to form a concentrated spray region with no gaps between the sprays. This was easily achieved due to the flexibility of the nozzle fixtures. This resulted in a concentrated spray pattern 100 mm long. A lower trolley speed of 300 m/s was then used to give the required quench time. The final setup is given in Table 4 (Sample QST2).

The microstructure obtained (Fig. 9) shows the tempered martensite layer progressing through bainite–ferrite to a ferrite–pearlite interior, typical of QST products. Therefore, by making only minor adjustments to the configuration of the sprays, the unit has successfully achieved the degree of control necessary to simulate more complex processes.

4.2 MULTISTAGE COOLING

To demonstrate the multistage cooling capability of the unit, a 27 mm thick, 100 mm long and 100 mm wide plate was rolled according to Schedule 2 in Table 1, followed by three-stage controlled cooling with the setup as outlined in Table 4. The aim was to rapidly cool the sample from the finish rolling temperature (850°C) to between 700 and 720°C, followed by a 10 s natural air cooling stage and then final rapid cooling to 500°C.

To achieve this cooling pattern each zone, comprising three banks of three top and bottom nozzles, was setup to reproduce a given stage, thus the number of sprays in the first bank and the trolley travel speed were adjusted to give the desired temperature drop for a single pass through that zone by the strip. The slow air cooling stage was obtained by turning off the second zone of sprays and holding the strip stationary in this position for the required time, after which the strip was passed through the final zone to produce the second rapid cooling phase. The results of the simulation (Fig. 10) shows that these complex rolling and cooling scenarios, where a combination of both controlled rolling and multi-stage cooling are necessary, can be successfully achieved on this facility in conjunction with the experimental mill.

5 CONCLUSIONS

A laboratory spray cooling unit has been constructed which is capable of simulating complex thermomechanical controlled processing and in-line heat treatment techniques. A wide range of cooling rates from 10°C/s to over 50°C/s has been achieved for 12 mm thick plate. A temperature model has been developed which accurately predicts the temperature history during the deformation, air cooling and forced spray cooling phases of the simulation. The heat transfer coefficient for the atomising air/water nozzles was found to be approximately three times greater than that expected for purely hydraulic nozzles with the same water flux. This model allows accurate planning of experiments which reduces the total number required to carry out a particular study. The cooling unit and adjacent rolling mill are well instrumented and work is continuing on the development of feed-forward and feed-back computer control of the cooling rig.

ACKNOWLEDGEMENTS

The authors would like to thank The Broken Hill Proprietary Co Ltd. for permission to publish this work. They are also grateful to M.S. Jenkins and S.R. Story for their detailed characterisation of the spray nozzle, and J. White and A. Deuchar for technical assistance during the rolling and cooling investigations.

REFERENCES

1. M.S. Jenkins, S.R. Story and R.H. Davies: 'Measurement and Characterisation of Air Mist Nozzles for Spray Quenching Heat Transfer', elsewhere in these proceedings.
2. P.D. Hodgson, J.A. Szalla and P.J. Campbell: *4th Int. Steel Rolling Conference, Deauville, France*, 1987, paper C8.
3. P.D. Hodgson, K.M. Browne, D.C. Collinson, T.T. Pham and R.K. Gibbs, elsewhere in these proceedings.
4. T.M. Casterjet: Bulletin 202, Spraying Systems Co. information leaflet.

13

Microstructure, Residual Stresses and Fatigue of Carburised Steels

G. KRAUSS

Advanced Steel Processing and Products Research Center, Colorado School of Mines, Golden, Colorado 80401, USA

ABSTRACT

This paper reviews the evolution of microstructure and residual stress during the heat treatment processing of gas carburised steels and relates these factors to fatigue resistance. The tempered martensite-retained austenite microstructure produced by quenching and tempering, and the prior microstructure produced by carburising at temperatures where austenite is stable are described. Austenising affects grain growth, segregation of phosphorus, carbide formation, and surface oxidation. Factors which affect the development of surface residual stresses, i.e. quenching, refrigeration, oxide formation, tempering, shot peening, and strain-induced transformation of austenite, are discussed. The microstructure and residual stresses are related to fatigue crack initiation mechanisms and fatigue performance.

1 INTRODUCTION

Carburised steels are widely used for shafts, gears, bearings and other highly stressed machine parts. Most of these machine components are cyclically stressed and therefore resistance to fatigue fracture is a critical design factor. Resistance to rolling contact fatigue and/or bending fatigue may be required, depending on the application. This paper is restricted to bending fatigue performance.

Many processing and microstructural factors affect bending fatigue performance, and wide variations in endurance limits, ranging from 210 MPa to 1950 MPa, have been reported in the literature.[1] Endurance limits in the mid and upper ranges of the reported values represent very good resistance to fatigue fracture, and are sensitive to microstructural factors such as the amounts and distribution of martensite and retained austenite and the development of favourable surface residual compressive stresses. The purpose of this paper is to review (a) the evolution of the processing-dependent microstructures and residual stresses in carburised steels, and (b) the relationship of microstructure and residual stress to fatigue performance of carburised steels.

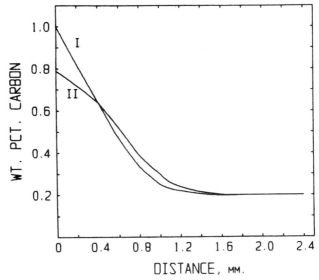

Fig. 1. Simulated carbon profiles for two stage gas carburising. Stage I: 927°C, 1.1 pct C potential, 3 hours. Stage II: 871°C, 0.8 pct C potential, 1.5 hours.[2]

2 PROCESSING AND MICROSTRUCTURE

2.1 PROCESSING CONSIDERATIONS

Carburising of low-carbon steel may be accomplished by many techniques including gas, vacuum, plasma, salt bath and pack carburising.[2] By far the most widely used process for large volume production is gas carburising. In this process, the steel is exposed to a carrier gas atmosphere with gaseous hydrocarbons and carbon monoxide which decompose to introduce carbon into the surface of a steel.[2,3] The steel is held at a temperature at which the microstructure is austenitic, typically 930°C, and carbon diffuses into the interior of the steel. Typically, the carburising is done in two stages: an initial stage during which the carbon potential is maintained at a level close to the solubility limit of carbon in austenite, at a carbon level which ranges from 1.0 to 1.2 pct C depending on temperature and alloy content of the steel, and a second step during which the carbon potential of the atmosphere is reduced to a level that will maintain surface carbon between 0.8 and 0.9 pct. In the second stage the excess case carbon diffuses further into the steel. The two-step process is therefore often referred to as boost-diffuse carburising. Figure 1 shows a computed example of carbon gradients produced by two stage carburising.[2] When the required case depth is achieved, the temperature is lowered to 850°C and the parts are quenched. Quenching from 850°C reduces distortion.

The austenite carbon gradients translate into hardness gradients after quenching. Figure 2 shows almost identical hardness gradients produced for the same steel, SAE 8719 containing 1.06 pct Mn, 0.52 pct Cr, 0.50 pct Ni, and 0.17 pct Mo, carburised and hardened by four different schedules.[4] Case depths are usually

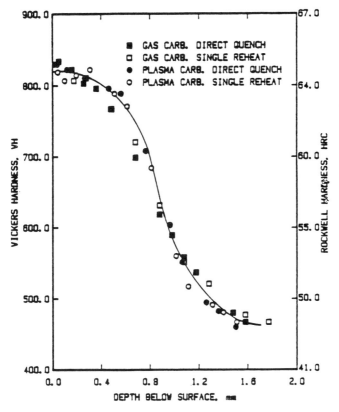

Fig. 2. Hardness profiles from SAE 8719 steel carburised and hardened as shown.[4]

defined as the distance from the surface to the point where hardness drops to a given level, typically HIV 510 or HRC 50.

2.2 AUSTENITE MICROSTRUCTURE

The microstructure during carburising consists of polycrystalline austenite. Since carburising times are relatively long, for example 4 to 5 hours to produce a case depth of 1 mm at 930°C, there is a possibility of grain coarsening during carburising. As a result carburising steels are almost universally fine-grained, aluminium-killed steels in which aluminium nitride particles suppress grain growth. Figure 3 shows an example of the austenite grain structure produced by reheating a carburised low-carbon Cr-Ni-Mo steel. The specimen has been etched to bring out the austenitic grain boundaries.[5] Austenite grain size determines the size and distribution of the martensite which forms during quenching, and therefore plays an indirect but important role in fatigue resistance of carburised steels.

Another very important structural change which develops in the austenitic

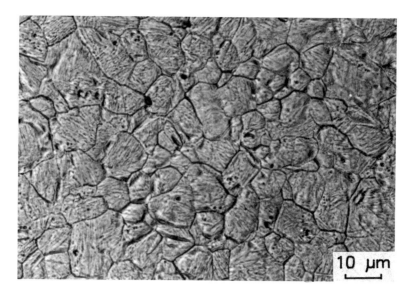

Fig. 3. Prior austenite grain structure in SAE 8719 steel gas carburised and reheated to 850°C.[4]

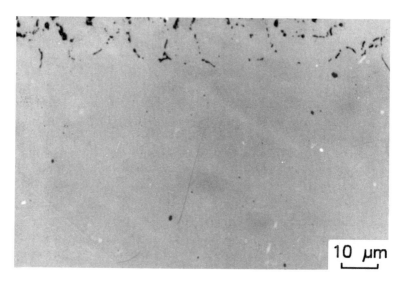

Fig. 4. Surface oxidation in gas-carburised SAE 8719 steel.[4]

structure during carburising is the segregation of phosphorus to austenite grain boundaries. The phosphorus concentrates in very thin layers, on the order of atomic dimensions, and can only be detected by Auger electron spectroscopy (AES). Such analysis has shown that phosphorus is present at austenitic grain boundaries in as-quenched steels.[6,7] Thus tempering is not required to cause the segregation.

During quenching, very small amounts of cementite form on the austenite grain boundaries in the high-carbon case, and the combination of phosphorus and cementite at the grain boundaries leads to a very high sensitivity to intergranular fracture.[7,8,9] Intergranular cracking, as discussed later, is a major cause of fatigue crack initiation in carburised steels.

Surface oxidation is another very important microstructural change introduced during gas carburising when the steel is in the austenitic state.[10,11] All gas carburised atmospheres contain some partial pressure of oxygen due to the presence of CO_2 and H_2O,[2] and this oxygen preferentially combines with some of the alloying elements present in the steel. Figure 4 shows an example of surface oxidation in a gas carburised SAE 8719 steel.[4] The surface oxidation has been found to develop two zones: a shallow surface zone in which chromium-rich oxides form within austenite grains, and a deeper zone in which manganese-rich and silicon-rich oxides form along austenite grain boundaries.

The presence of oxides in the near surface zones of gas-carburised steels raises

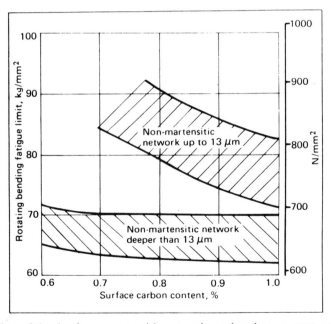

Fig. 5. Effect of depth of non-martensitic networks and carbon content on fatigue of carburised steels.[12]

questions regarding the effect of oxides on fatigue fracture. There are two aspects which cause concern. One is the effect of alloying element depletion by the oxide formation to the point where hardenability is decreased and transformation products other than martensite form in surface layers. The other is the role that the oxides play in the initiation of fatigue cracks. The first aspect affects the residual stresses developed in the case and will be discussed later. The second aspect is related to the depth of the surface oxide zones. The depth of oxide penetration is diffusion-controlled and therefore dependent on the time of carburisation. Figure 5 shows results of experiments which show that deeper non-martensite networks related to internal oxidation more severely degrade fatigue resistance than do shallow networks.[12] More recent work shows that very good fatigue resistance can be achieved with oxide penetration on the order of 10 μm when the surface microstructure associated with the oxidation remains martensitic.[4,13]

Excess carbon, above the amount required for carbide formation, may be introduced into austenite during carburising. If this excessive carbon content is not lowered during the diffusion stage of a carburising cycle, massive carbides will form either during carburising or during cooling.[12] Severe massive carbide formation has been associated with high temperature carburising during which the carbon solubility in austenite in equilibrium with cementite is very high. The carbides form at specimen corners where lowering of carbon content by diffusion is geometrically constrained, and serve as preferred sites for fatigue crack initiation.[14]

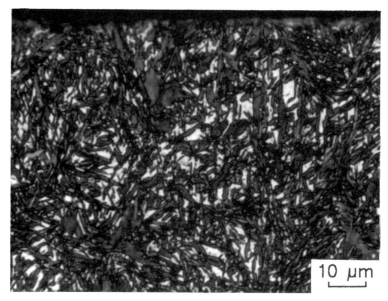

Fig. 6. Plate martensite and retained austenite in case of carburised and direct quenched SAE 4121 steel.[14]

2.3 MARTENSITE–AUSTENITE–CARBIDE MICROSTRUCTURES

When the austenite with the surface carbon gradients introduced by carburising is quenched, the austenite transforms to martensite. In the high-carbon case the martensite forms in a plate morphology, and because of the low M_s temperatures associated with the high carbon content significant volume fractions of austenite are retained.[15] In the low-carbon core, martensite forms in a lath morphology with no resolvable austenite. As the carbon content drops from the surface to the core there is a transition from plate to lath martensite morphology, and the amount of retained austenite decreases continuously. Figures 6 and 7 show examples of plate and lath martensite which have formed in the case and core regions of carburised specimens, respectively.

The formation of martensitic structures in carburised steels is a function of hardenability.[16] Alloy carburising steels are selected to insure good case and core hardenability, but depending on alloying, quenching factors and section size, varying amounts of non-martensitic transformation products of austenite may form in the case and core.

Figure 6 shows the martensite typically formed in the case of steels directly quenched after carburising. If carburised steels are reheated to a temperature below A_{cm}, i.e. at a temperature where austenite and cementite are stable, distributions of spheroidised carbide particles are developed in the case as shown in Fig. 8. The carbides tie up some of the carbon, and therefore M_s increases, producing less retained austenite in quenched specimens. Also the carbide particles

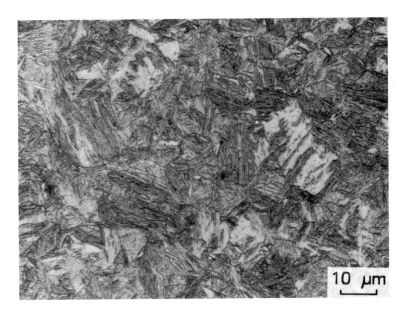

Fig. 7. Lath martensite in core of carburised and re-heated SAE 8719 steel.[4]

212 *Quenching and Carburising*

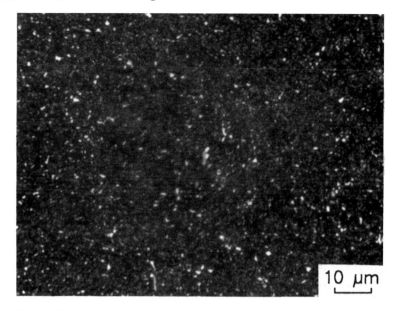

Fig. 8. Spheroidised carbides (white) in matrix of martensite (dark) of carburised and reheated SAE 8620 steel.[14]

effectively reduce grain boundary migration, and therefore, depending on their volume fraction and size, maintain a very fine austenite grain size.

Almost all carburised steels are tempered, generally at temperatures between 150°C and 200°C. These low tempering temperatures preserve high strength and hardness. Residual stresses are reduced, fine transition carbides precipitate within martensite plates, and there is no change in austenite content from that retained in the as-quenched condition as a result of low-temperature tempering treatments.[15]

3 RESIDUAL STRESSES

3.1 GENERAL CONSIDERATIONS

The chemical and transformation gradients in carburised steels produce surface residual compressive stresses. These stresses make it possible to readily manufacture high carbon martensitic structures which in quenched, through-hardened steels are difficult to produce because of surface tensile stresses and high sensitivity to quench cracking.[15,17] Also, the surface compressive stresses in carburised steels offset applied tensile bending stresses and thereby significantly increase fracture and fatigue resistance.

The compressive surface stresses are a result of the effect of temperature gradients established during quenching superimposed on M_s gradients associated with the carbon gradients introduced into the austenite by carburising.[18] At some point during quenching, the temperature falls below the higher M_s temperature in the interior of the specimen, and the austenite begins to transform to martensite. Although the temperature is lower at the surface, the austenite there remains untransformed because of the low surface M_s. The interior formation of martensite causes an associated volume expansion which at high temperatures can be readily accommodated by the surrounding austenite. Eventually, during the quenching process, the temperature at the surface falls below the M_s and high carbon martensite forms. The interior martensite, now at a lower temperature and significantly stronger than austenite, resists the expansion of the high-carbon surface martensite and thereby places it in compression. Concomitantly the interior martensite is placed in tension.

Figure 9 shows a schematic diagram of the residual stress profiles that typically form in carburised and hardened steels.[19] The compressive residual stress peaks at some distance from the surface and the surface residual stresses are balanced by interior tensile residual stresses. Also shown schematically is the carbon profile responsible for the transformation gradients on quenching and a retained austenite profile. Larger amounts of austenite transformation would be expected to increase compressive stresses according to the scenario outlined above, but there is a limit to the benefits of austenite transformation, especially if accomplished by refrigeration treatments. The latter is discussed in a later section.

Parrish and Harper[19] report the results of a literature survey which included the evaluation of 70 residual stress curves for carburised steels. The specimens had case depths around 1 mm or less, the core carbon contents were between 0.15 and 0.20

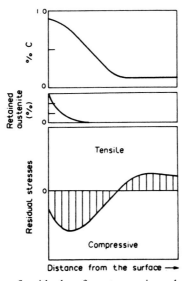

Fig. 9. Schematic diagram of residual surface stresses in carburised steels.[19]

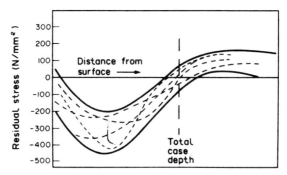

Fig. 10. Ranges of residual stresses measured in 70 carburised steels.[19]

pct, and the parts were oil quenched and tempered between 150 and 180°C. Figure 10 shows the range of residual stress profiles which are typical of gas carburising accomplished according to the conditions outlined above. As shown, the compressive residual stress zone encompasses most of the case.

3.2 MEASUREMENT AND MODELLING OF RESIDUAL STRESSES

Residual stress profiles are routinely measured by X-ray diffraction analysis.[20] The residual stresses cause changes in the interplanar spacings of the lattice planes in the martensite and austenite of carburised steels, and the changes in interplanar spacing are used to calculate strains which in turn are used to calculate stresses by use of Poisson's ratio and the Young's modulus.[21] Because of the limited depth of penetration of X-rays, residual stress profiles must be obtained by serial examination of subsurface layers exposed by electrolytic or chemical polishing.

The development of residual stresses depends on complex interactions between specimen size and geometry, heat flow through the steel, heat transfer associated with quenching, the transformation of austenite as a function of chemical composition, position and time as determined by cooling rates and the composition gradients established by carburising, and the temperature-dependent mechanical properties and plastic flow characteristics of the mixtures of austenite, martensite and other phases which form as a carburised specimen is quenched. All of these phenomena are now being incorporated into computer models which predict residual stresses and distortion of carburised steels, and although a major shortcoming is the unavailability of accurate high-temperature constitutive flow stresses for austenite, martensite and other phases, the models make valuable predictions of residual stress as a function of variations in processing parameters.[22-26]

Table 1 lists a series of residual stress parameters calculated by Ericsson et al.[23] for cylinders of various sizes of a low-carbon steel carburised to three different case depths. Calculated and measured values agree well, and, with increasing bar

Microstructure, Residual Stresses and Fatigue of Carburised Steels

Table 1. Compilation of characteristic depths and residual stresses for case hardened cylinders[23]

Diam. (mm)	Case depth DC (mm) calc.	Case depth DC (mm) exp.	σ_{max} MPa calc.	Depth to σ_{max} mm calc.	Depth to σ_{max} divided by DC calc.	Depth to start mart mm calc.	Depth to 0-crossing mm calc.	Core hard HVI exp.
10	0.67	0.67	−450	0.06	0.10	1.0	0.83	360
17	0.66	0.63	−520	0.15	0.23	0.9	0.87	320
30	–	–	−540	0.15	–	–	0.86	–
10	0.98	1.04	−420	0.06	0.06	1.7	1.18	360
17	0.95	0.94	−480	0.25	0.27	1.3	1.40	320
30	0.90	0.83	−520	0.45	0.50	1.1	1.41	280
10	–	–	−340	0.25	–	–	1.68	–
17	1.89	1.74	−415	0.60	0.32	1.3	2.10	380
30	1.71	1.49	−480	0.90	0.53	1.1	2.35	260

diameter, case depth decreases because of increased formation of non-martensitic products. However, the calculations show that the maximum compressive stresses and the position of the maximum increase with increasing bar diameter. This finding is related to the greater amount of non-martensitic transformation products in the core of the heavier sections. The latter situation is explained by a greater difference in the case transformation strain and the average transformation strain, leading to the predicted increased compression residual stress.

3.3 RESIDUAL STRESSES AND SURFACE OXIDATION

In another study, Hildenwall and Ericsson[24] calculate the tensile stresses produced by excessive surface oxidation produced by gas carburising. If the oxidation is severe enough, sufficient alloying elements are removed from solid solution and the hardenability of the surface may be decreased to the point where pearlite forms instead of martensite. As a result the surface transforms at high temperature early in the quenching cycle and when the unoxidised case austenite eventually transforms to martensite, the resulting expansion places the surface in tension. The surface residual tensile stresses then adversely affect fatigue behaviour.[11,12,27]

3.4 RESIDUAL STRESSES AND SHOT PEENING

Systematic measurements of residual stress profiles provide valuable information concerning the performance of carburised steels. Sometimes a direct correlation can be made to fatigue performance if parallel fatigue testing is performed. In other cases, if high values of residual compressive stresses are measured, improved fatigue performance can be inferred. For example, shot peening is commonly

216   *Quenching and Carburising*

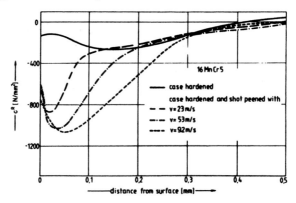

Fig. 11. Effect of shot peening at velocities marked on residual stresses.

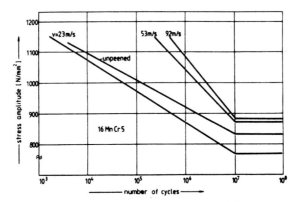

Fig. 12. Effect of shot peening on fatigue of carburised steel.[19]

applied to carburised steels in order to increase fatigue performance. Figure 11 shows significant increases in residual surface compressive stresses due to increased shot peening velocity, and Fig. 12 shows the improved fatigue performance produced by the shot peening.[21] Kim et al.[28] also measured the effects of shot peening on residual stress and showed a beneficial effect on fatigue performance.

3.5 QUENCHING AND TEMPERING EFFECTS ON RESIDUAL STRESS

Shea[29,30] has systematically measured the effect of quenchant temperature and tempering on the residual stresses, microstructure and properties of carburised steels. Figures 13 and 14 show the effect of quenchant temperature and tempering on the surface compressive stresses of SAE 4130, containing 1.0 Cr and 0.20 pct Mo, and SAE 1526 containing 1.25 pct Mn, respectively. The compressive residual

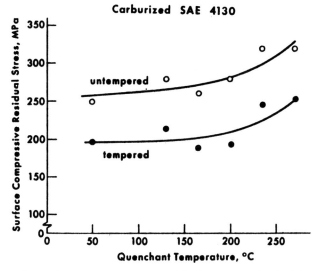

Fig. 13. Effect of quenchant temperatures on residual stress in carburised SAE 4130 steel.[29]

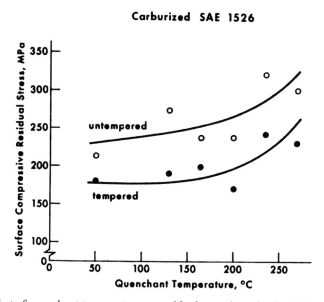

Fig. 14. Effect of quenchant temperature on residual stress in carburised SAE 1526 steel.[29]

stresses increase with increasing quenchant temperature from 50°C typical of conventional warm oil quenches to 270°C which is classified as a hot oil quench. Tempering at 165°C in all cases reduced the surface compressive stresses. The high residual stresses in the hot oil quench of the high-hardenability 4130 steel were

218 *Quenching and Carburising*

attributed to a reduced temperature gradient which shifted the transformation start temperature toward the core, causing substantial core transformation prior to the surface transformation. The lower compressive stresses of the hot oil quenched low-hardenability SAE 1526 steel were attributed to lower transformation strains associated with bainite rather than martensite.

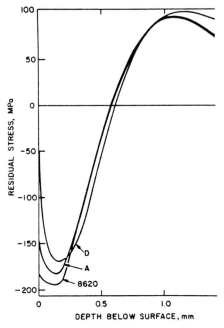

Fig. 15. Residual stresses in gas carburised steels quench and tempered at 150°C. Steels A and D are secondary hardening steels.[31]

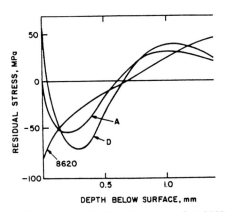

Fig. 16. Residual stresses in gas carburised steels tempered at 300°C. Steels A and D are secondary hardening steels.[31]

A study by Stickels and Mack[31] also evaluated the effect of tempering on surface residual stress formation in carburised steels. They examined carburised Mo-Cr-V secondary hardening steels and a carburised SAE 8620 steel, a commonly used low-alloy carburising steel. Figures 15 and 16 show residual stress profiles after tempering at 150°C and 300°C. At 300°C, most of the retained austenite has transformed and the martensitic structure is decreasing in volume as the tetragonality of the martensite is relieved by carbide formation.

3.6 RESIDUAL STRESS AND REFRIGERATION

Refrigeration or sub-zero cooling treatments are sometimes applied to carburised steels. The lower temperatures cause some of the retained austenite to transform to martensite, and thereby increase case hardness especially if excessive amounts of austenite were present at room temperature. Reduction in retained austenite content also increases dimensional stability by minimising the volume expansion associated with thermally strain-induced transformation of austenite to martensite during tempering or in service. However, a number of investigators show that refrigeration treatments are detrimental to fatigue resistance.[28,32,33]

The reason for the decreased fatigue performance has been shown to be associated with localised residual stresses developed by refrigeration treatments. Although compressive stresses continue to develop in the martensite of the case, Kim et al.[28] have measured high tensile stresses in the austenite of sub-zero cooled specimens. These tensile stresses then act in concert with applied bending tensile stressed to cause crack initiation at susceptible microstructural features during cyclic loading. If refrigeration is used, the carburised parts should be tempered before and after sub-zero cooling.[19]

3.7 STRAIN-INDUCED AUSTENITE TRANSFORMATION AND RESIDUAL STRESS

The deformation-induced transformation of austenite to martensite is well known in stainless and other highly alloyed steels.[34] The relatively large amounts of retained austenite, typically between 20 and 30 pct, in the near surface case regions of direct-quenched carburised steels also make deformation-induced austenite transformation a factor in the fatigue performance of carburised steels. Zaccone et al.,[35] in a study of the bending fatigue of notched specimens of a series of high carbon steels with microstructures similar to those in the case of carburised steels, show that significant amounts of austenite transform in the plastic zone around a growing fatigue crack. The strain-induced austenite transformation was shown to be beneficial to low-cycle, high strain fatigue resistance. Specimens with more austenite showed longer fatigue life than specimens with lower amounts of retained austenite. The beneficial effects of the strain-induced austenite transformation were attributed to the associated increased strain hardening and generation of

220 Quenching and Carburising

increased compressive stresses which arrested and slowed fatigue crack propagation.

Shea, in a study of impact properties of carburised steels, has measured the changes in surface residual stresses which develop as a function of bending strain in carburised steels.[36] Figures 17 and 18 show results for carburised 4120 and 3310 steels, respectively. In the as-quenched and tempered specimens, containing greater than 20 pct retained austenite after an incubation strain, compressive stresses increase significantly with strain. These increases in compressive stress correlated directly with strain-induced austenite transformation. Specimens quenched in liquid nitrogen, with much reduced retained austenite contents, about 10 pct when strained, showed a slight decrease on compressive stress. Higher fracture resistance was associated with the strain-induced martensite formation in the microstructures with the larger amounts of retained austenite.

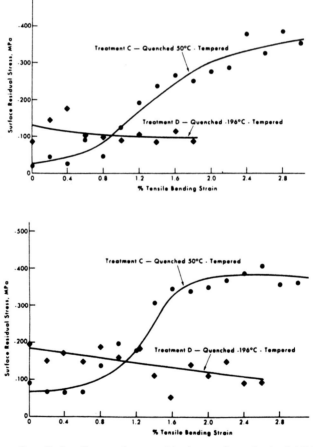

Fig. 17. Effect of tensile bending strain on residual stress in carburised 4620 steel.[36]

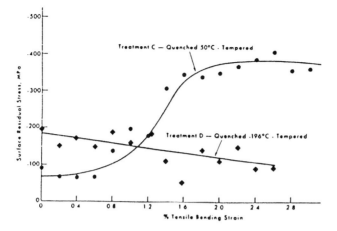

Fig. 18. Effect of tensile bending strain on residual stress of carburised 3310 steel.[36]

4 FATIGUE MECHANISMS AND PERFORMANCE

Fatigue cracks in carburised steels subjected to cyclic bending stresses may be nucleated at a variety of features: surface discontinuities such as machining grooves and pits, inclusions, massive carbides and prior austenite grain boundaries. Fatigue crack growth is limited because of the relatively low case fracture toughness, 20 to 27 MPam$^{1/2}$, of the high carbon martensitic microstructure.[37] Once the critical crack size is achieved, unstable crack propagation proceeds through the balance of the case and into the core.

Fig. 19. Macrograph of fatigue fracture cross section of carburised Type 8219 steel. Granular fracture of case and ductile fracture of core is shown.[13]

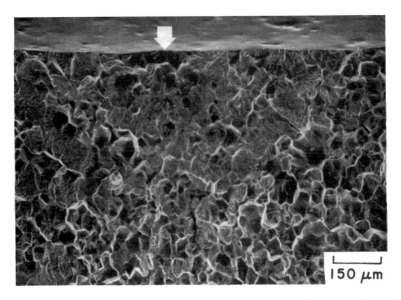

Fig. 20. Fatigue crack initiation (arrow), transgranular growth zone, and overload intergranular zone in simulated carburised case microstructure in an 0.85 pct C steel.[35]

Fig. 21. Intergranular fatigue crack initiation (arrow) and transgranular crack propagation zone in a simulated carburised case microstructure in an 0.85 pct C steel.[35]

Figure 19 shows the macroscopic fracture cross-section through a bending fatigue specimen of a carburised Type 8219 steel.[13] The case fracture region has a granular appearance and is delineated from the core fracture which has a ductile, fibrous appearance. On this scale the fatigue crack initiation and propagation areas are too small to be resolved.

Figures 20 and 21 show the fatigue crack initiation and growth areas at higher magnifications. These fractographs were taken from a high carbon steel with a plate martensite-retained austenite microstructure similar to that in the near surface case region of carburised steels. Carburised steels subjected to bending fatigue show exactly the same type of fatigue crack topography.[4,13,35] The arrows in Figs. 20 and 21 point to intergranular fracture initiation sites. These first cracks are arrested, and slow transgranular fatigue crack growth proceeds. When the critical crack size is attained, unstable crack propagation occurs by intergranular fracture as best shown in Fig. 20. The latter intergranular fracture makes up the case fracture shown in the macrograph of Fig. 19.

The above sequence of fatigue fracture is typical of that observed in properly carburised specimens, i.e. those without excessive surface oxidation or massive carbides, with smooth surfaces. The intergranular crack initiation and intergranular unstable case crack propagation are caused by the combination of phosphorus segregation and cementite formation at austenite grain boundaries as described earlier. The intergranular crack initiation extends only a few grains, and is

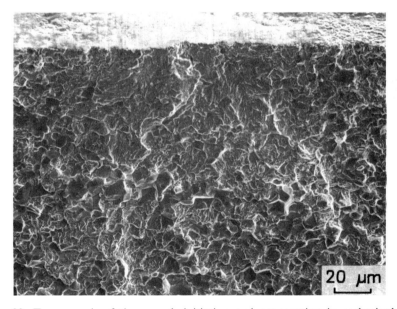

Fig. 22. Transgranular fatigue crack initiation and propagation in carburised and reheated SAE 8719 steel.[24]

arrested, perhaps because of the generation of favourable residual compressive stresses by strain-induced retained austenite transformation.[35,36]

Erven[13] has shown in a series of eight alloy steels that the intergranular initiation mode of fatigue cracking is associated with endurance limits between 1070 MPa and 1260 MPa. This range of fatigue resistance appears to be the range associated with well-prepared surfaces and small ranges in residual stresses, austenitic grain size, alloy content and retained austenite. Rough surfaces, massive carbides, excessive surface oxidation and high densities of inclusions would all lower fatigue performance.

A second mechanism of fatigue cracking in carburised steels is associated with transgranular crack initiation as shown in Fig. 22. These cracks indicate at surface discontinuities and apparently depend on slip mechanisms of fatigue crack initiation.[38] The endurance limits associated with this type of initiation are much higher, 1400 to 1950 MPa, than those associated with microstructures sensitive to intergranular cracking. Carburised microstructures which have these high levels of fatigue resistance have very fine austenitic grain sizes and relatively low retained austenite contents, microstructures which may be produced by reheating carburised specimens to below A_{cm}.[4,39]

ACKNOWLEDGEMENTS

I thank Ms Mimi Martin for her assistance with the preparation of this manuscript and Scott Hyde for assistance with the figures. Research on fatigue of carburised steels at the Colorado School of Mines is supported by the Advanced Steel Processing and Products Research Center.

REFERENCES

1. R.E. Cohen, P.J. Haagensen, D.K. Matlock and G. Krauss: SAE Technical Paper No. 910140, SAE, Warrendale, PA, 1991.
2. C.A. Stickels and C.M. Mack, in *Carburizing: Processing and Performance*, ed. G. Krauss, ASM International, Materials Park, OH, 1989, p. 1.
3. *Case Hardening of Steel*, H.E. Boyer ed., ASM International, Materials Park, OH, 1987.
4. J.L. Pacheco and G. Krauss: *J. Heat Treating, 1989*, **7**, (2), 77.
5. A. Brewer, K.A. Erven, and G. Krauss, to be published in *Materials Characterization*.
6. O. Ohtani and C.J. McMahon, Jr: *Acta Metallurgica, 1975*, **23**, 337.
7. T. Ando and G. Krauss: *Metall. Trans. A, 1981*, **12A**, 1283.
8. G. Krauss: *Metall. Trans. A, 1978*, **9A**, 1527.
9. H.K. Obermeyer and G. Krauss: *J. Heat Treating, 1980*, **1**, (3) 31.
10. R. Chatterjee-Fischer: *Metall. Trans. A, 1978*, **9A**, 1553.
11. C. Van Thyne and G. Krauss, in *Carburizing: Processing and Performance*, G. Krauss, ASM International, Materials Park, OH, 1989, p. 333.
12. G. Parrish, *The Influence of Microstructure on the Properties of Case-Carburized Components*, ASM, Materials Park, OH, 1980.

13. K.A. Erven, D.K. Matlock and G. Krauss: *Materials Science Forum*, 1992, 102–4, 183.
14. K.D. Jones and G. Krauss, in *Heat Treatment '79*, The Metals Society, London, 1979, 188.
15. G. Krauss: *Steels: Heat Treatment and Processing Principles*, ASM International, Materials Park, OH, 1990.
16. U. Wyss: *Harterei Technisehe Mitteilungen*, 1988, **43**, 27.
17. L.J. Ebert: *Metall. Trans. A*, 1978, **9A**, 1537.
18. D.P. Koistinen: *Trans. ASM*, 1958, **50**, 227.
19. G. Parrish and G.S. Harper: *Production Gas Carburizing*, Pergamon Press, New York, 1985.
20. *SAE Handbook Supplement: Residual Stress Measurement by X-Ray Diffraction*, SAE J784a, SAE, Warrendale, PA, 1971, p. 19.
21. G. Scholtes and E. Macherauch, in *Case-Hardened Steels: Microstructural and Residual Stress Effects*, by D.E. Diesburg ed., TMS, Warrendale, PA, 1984, p. 141.
22. B. Hildenwall and T. Ericssen, in *Hardenability Concepts with Applications to Steel*, D.V. Doane and J.S. Kirkaldy eds., TMS, Warrendale, PA, 1978, p. 579.
23. T. Ericsson, S. Sjostrom, M. Knuuttila and B. Hildenwall, in *Case-Hardened Steels: Microstructural and Residual Stress Effects*, D.E. Diesburg ed., TMS, Warrendale, PA, 1984, p. 113.
24. B. Hildenwall and T. Ericsson: *J. Heat Treating*, 1980, **1**, (3), 3.
25. J.A. Burnett, in *Residual Stress for Designers and Metallurgists*, L.J. Vande Walle ed., ASM, Materials Park, OH, 1981, p. 51.
26. M. Henriksen, D.B. Larson and C.J. Van Tyne: *Trans. A.S.M.E.*, 1992, **114**, 362.
27. S. Gunnarson: *Metal Treatment and Drop Forging*, 1963, **30**, 219.
28. C. Kim, D.E. Diesburg and R.M. Buck: *J. Heat Treatment*, 1981, **2**, (1), 43.
29. M.M. Shea: *J. Heat Treating*, 1980, **1**, (4), 29.
30. M.M. Shea: *J. Heat Treating*, 1983, **3**, (1), 38.
31. C.A. Stickels and C.M. Mack: *J. Heat Treating*, 1986, **4**, (30), 223.
32. K.D. Jones and G. Krauss: *J. Heat Treating*, 1979, **1**, (1), 64.
33. M.A. Panhans and R.A. Fournelle: *J. Heat Treating*, 1981, **2**, (1), 55.
34. G.B. Olson and M. Cohen: *Metall. Trans. A*, 1975, **6A**, 791.
35. M.A. Zaccone, J.B. Kelley and G. Krauss, in *Carburizing: Processing and Performance*, G. Krauss ed., ASM International, Materials Park, OH, 1989, 249.
36. M.M. Shea: SAE Technical Paper No. 780772, SAE, Warrendale, PA, 1978.
37. G. Krauss, in *Case-Hardened Steels: Microstructural and Residual Stress Effects*, D.E. Diesburg ed., TMS, Warrendale, PA, 1984, 33.
38. *Fatigue and Microstructure*, M. Meshii ed., ASM, Metals Park, OH, 1979.
39. C.A. Apple and G. Krauss: *Metall. Trans.*, 1973, **r**, 1195.
40. L. Magnusson and T. Ericsson, in *Heat Treatment '79*, The Metals Society, 1980, p. 202.

14
Fundamentals of Carburising and Toughness of Carburised Components

J. GROSCH

Institut für Werkstofftechnik, Technische Universität, Berlin, Germany

ABSTRACT

Fundamental principles of gas carburising, the most widely used carburising process in industry, are discussed, such as gas equilibria, carbon diffusion, kinetics and controlling. Characteristic carburised microstructures are derived from heat treatment parameters and shown in optical micrographs. The influences of the case depths in general and the microstructure of the case in particular on toughness and ductility of carburised components are described.

1 INTRODUCTION

Carburising produces a hard and – compared with their dimensions – often a small surface on relatively soft components when the surface microstructure of steels with a (core) carbon content, usually of 0.15–0.25%, is carburised to carbon contents in the range 0.7–0.9 (1.0) %C and transformed to martensite. This treatment leads to the formation of a hardness gradient and to a distribution of residual stresses with compressive stresses in the surface microstructure due to the changes in volume during the martensitic transformation. The combined effect of these two parameters causes the main properties of the components, i.e. fatigue, rolling contact fatigue and toughness, to reach the highest values possible in *one* part.

The process of carburising thus consists of two steps: *carbon diffusion*, with the result of a carbon gradient, and *quenching*, with the hardness gradient and the resulting distribution of residual stresses as results. While it is possible to establish equal carbon gradients in different alloy steels (Fig. 1),[1] it is not always possible to establish equal hardness gradients (Fig. 2)[2] because hardenability varies with the alloy composition of the steels.

The following discussion describes fundamental principles of carburising, looks

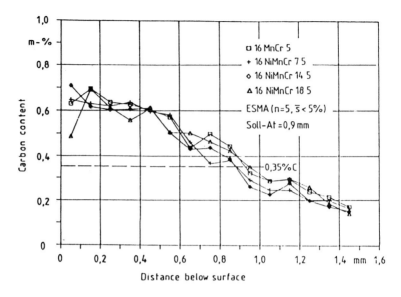

Fig. 1. Carbon gradients, computer controlled carburised.[1]

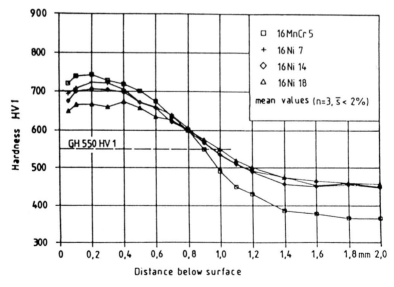

Fig. 2. Hardness gradients of steels with carbon gradients according to Fig. 1.[1]

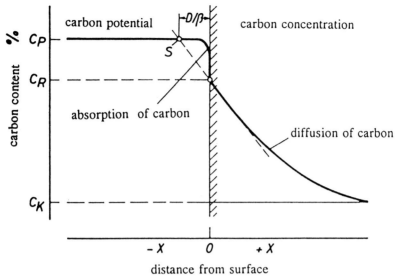

Fig. 3. Diagram of carburising process.[2]

into microstructures after quenching and also considers their toughness properties. Development and consequences of residual stresses as well as fatigue properties of carburised components are topics considered in Chapter 13.

2 FUNDAMENTALS OF CARBURISING

In the carburising process, steel is annealed at temperatures in the homogeneous austenite phase field in an environment of appropriate carbon sources; the carburising time depends on the desired case depth. The industrially used carburising processes are named after their carbon sources: *pack carburising* (solid compounds), *salt bath carburising* (liquid carbon sources) and *gas* and *plasma carburising* (gaseous carbon sources). The following discussion focuses on gas carburising, which is the dominant industrial process to date.

The process of gas carburising can be described by two steps:

- reactions in the gaseous atmosphere and absorption of carbon at the boundary layer at the surface of the steel;
- diffusion of carbon from the surface to the interior of the steel.

These are shown schematically in Fig. 3.[2]

The reactions in the gaseous atmosphere and the absorption of carbon at the surface can be described as gas equilibria;[3–8] the enforced non-equilibrium process leads to carburising. A consideration such as this is allowed since in the technical performance of the carburising process the diffusion of carbon in the steel, i.e. step two, determines the rate of carburising and is considerably slower than the other carburising step.

2.1 EQUILIBRIUM GAS COMPOSITION

Gaseous carbon sources are mainly hydrocarbon gases such as methane or propane,[6-8] 4–6 alcohols and alcohol derivatives (above all methanol[9] and other organic carbon compounds.[10]

These carbon sources decompose at carburising temperature into the constituents carbon monoxide (CO) and hydrogen (H_2) with small amounts of carbon dioxide (CO_2), water vapour (H_2O), oxygen (O_2) and methane (CH_4). In general, certain constant proportions of carbon monoxide and hydrogen diluted with nitrogen (N_2) are used as a carrier gas, which can be produced separately in a gas generator or in the furnace directly. Such a carrier gas performs with the reactions

$$2\ CO = C + CO_2 \quad \text{Boudouard reaction} \tag{1}$$

$$CO + H_2 = C + H_2O \tag{2}$$

$$CO = C + \tfrac{1}{2}O_2 \tag{3}$$

$$CH_4 = C + 2H_2 \tag{4}$$

to introduce carbon into the surface of a steel. With the equilibrium reaction of the constituents of the atmosphere (water–gas equilibrium).

$$CO_2 + H_2 = CO + H_2O \tag{5}$$

the carbon dioxide from reaction (1) is decomposed again. These reactions have both a carburizing and a decarburizing effect, which means that after sufficient carburizing time an equilibrium between carburization and decarburization is established whose carbon content is known as carbon potential C_p[11] (it is customary to define the carbon potential as the carbon content of pure iron which is in equilibrium with the furnace atmosphere), which can be measured by shim stock analysis.[12]

Figure 4[5] shows the equilibrium between carburisation and decarburisation for the Boudouard reaction in the presence of pure carbon dependent on pressure and temperature. The equilibrium can be determined by means of the equilibrium constant

$$\lg K_b = \lg (p_{CO}^2/p_{CO_2} \cdot p_c^0) = \lg (p_{CO}^2/p_{CO_2}) \tag{6}$$

with the concentration of the gas constituents as the respective partial pressures and the concentration of carbon as vapour pressure p_c^0 of pure carbon (equal to unity). The equilibrium curves from Fig. 4 at $p_{abs} = 0.2$ bar would represent an amount of 20% carbon monoxide in the carrier gas.

In carburising steel the influence of the carbon content and the alloy composition on the carbon concentration must be considered by means of the activity of carbon a_c,[6] which is defined as the ratio of the partial pressure p_c of carbon in the given condition to the vapour pressure p_c^0 of pure carbon

$$a_c = p_c/p_c^0 \tag{7}$$

The activity of carbon for plain carbon steels can be calculated[4,11] according to

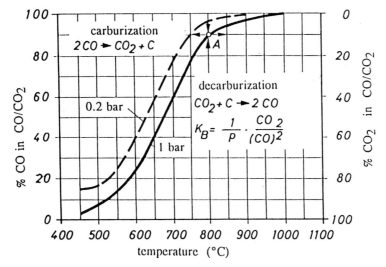

Fig. 4. Boudouard reaction in equilibrium with pure carbon.[5]

$$\lg a_c = 10\,500/4.575 \cdot T - (3.95 - 0.69 \cdot \%C)/4.575$$
$$+ \%C/(0.785 \cdot \%C + 21.5) \tag{8}$$

or, less complicated but sufficiently precise, as follows:

$$\lg a_c = 2300/T - 2.21 + 0.15 \cdot \%C + \lg \%C \tag{9}$$

as shown in Fig. 5.[13,14]

The influence of alloy composition on the activity of carbon is compensated by alloy factors,[6,8,15,16] which describe the ratio of the carbon potential of pure carbon to that of alloy steel and are used for calculating the required carbon potential of a carburising atmosphere.[17]

The equilibrium constant of the Boudouard reaction in the presence of steel[6,7,13] is

$$\lg k_B = \lg(p_{CO}^2/p_{CO_2} \cdot a_C) \tag{10}$$
$$= -8817/T + 9.071$$

(values according to Ref. 13, different values in Refs. 6,7) which therefore contains the activity of carbon, which can be determined according to:

$$\lg a_C = \lg(p_{CO}^2/p_{CO_2}) + 8817/T - 9.071 \tag{11}$$

The carburising equilibria in the homogeneous austenite phase field connected with the Boudouard reaction are shown in Fig. 6.[6] Similar equations and results are known for the reactions in equations (2)–(4), e.g. in Refs. 6, 13 and 18.

Given a certain composition and volume of gas, an equilibrium is eventually reached between the described reactions and a particular carbon potential, without further reactions in total. Methane decomposes with the reactions

$$CH_4 + CO_2 \rightarrow 2CO + 2H_2 \tag{12}$$
$$CH_4 + H_2O \rightarrow CO + 3H_2 \tag{13}$$

$$\lg a_C = \frac{2300}{T} - 2{,}21 + 0{,}15 \cdot (\%C) + \lg(\%C)$$

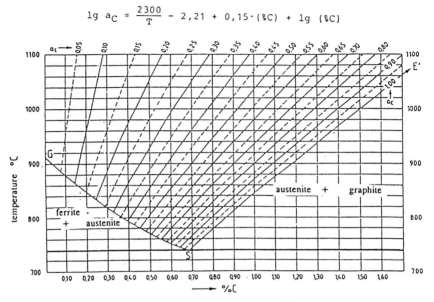

Fig. 5. Isoactivity of carbon in the austenite phase field.[13,14]

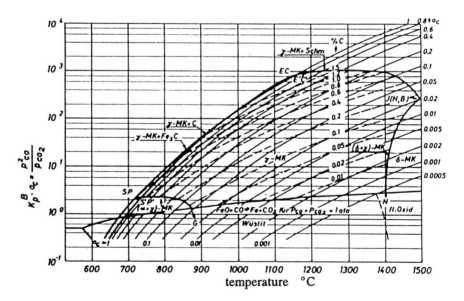

Fig. 6. Boudouard reaction in equilibrium with steel.[6]

the products carbon dioxide and water vapour of the carburising reactions (1) and (2), which leads to the regeneration of the carburising atmosphere. The reaction rates with methane according to equations (12) and (13), compared with the carburising reactions (1)–(3) and in particular with reaction (2), are slow, which means that rather large amounts of methane (or propane, which decomposes to methane) must be added to maintain the carburising process.

2.2 DIFFUSION OF CARBON IN STEELS

The diffusion of carbon follows Fick's laws[19]

$$\partial c/\partial t = -D \cdot \partial c/\partial x \tag{14}$$

and

$$\partial c/\partial t = -\partial/\partial x \cdot D \cdot \partial c/\partial x \qquad D = f(c) \tag{15a}$$

$$\partial c/\partial t = -D \cdot (\partial_c^2/\partial x^2) \qquad D = f(c) \tag{15b}$$

where
- c = concentration,
- t = time,
- x = distance,
- D = diffusion coefficient.

The Van-Ostrand–Dewey solution of equation (15a)[21,22]

$$(C(x,t) - C_0) = (C_R - C_0)(1 - \mathrm{erf}(x/2\sqrt{D \cdot t})) \tag{16}$$

describes the relationship between the carbon content $C(x,t)$ at a distance x from the surface of a steel with the initial carbon content C_0 after a carburising time t, assuming that the carbon content at the surface C_R is equal to the carbon potential (erf: error function). More precise calculations such as in Refs. 22–27, which take into account the differences between the carbon potential and the respective carbon content at the surface of the steel as well as the influence of the carbon concentration on the diffusion coefficient, serve as a basis for computer controlled carburisation, which was introduced after appropriate control systems[28,29] and computers[30] were available.

The diffusion of carbon determines the rate of carburising, for modern equipment offers enough carbon for absorption at the surface. Assuming this, the carburising time can only be reduced by higher temperatures[31,32] or/and by a steeper gradient of carbon concentration, which is achieved by computer controlled carburising.[1,33]

2.3 KINETICS OF CARBURISING

In carburising a carrier gas of a composition as constant as possible is enriched with an additional hydrocarbon, preferably methane or propane, to establish and to maintain a suitable carbon potential, which of course must exceed the carbon

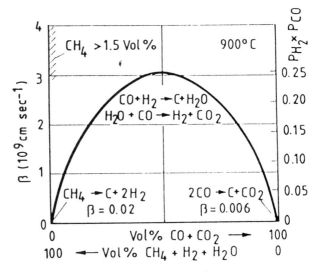

Fig. 7. Surface rate constant versus gas composition.[7]

content of the steel. The described equilibria are shifted to a *non-equilibrium condition*[34] in which the activity of carbon in the atmosphere, a_{cg}, is higher than the activity of carbon in the steel, a_{cs}. The difference in the activities acts as the driving force to transfer carbon from the atmosphere into the steel, with the surface flux $\dot{m}$ (number of atoms transferred in the time t through the area F) being proportional to the difference in activities[11]

$$\dot{m} = M/F \cdot dt = -\beta(a_{CG} - a_{CS}) \qquad (17)$$

The surface reaction rate constant β is in particular dependent on the composition of the atmosphere and on the temperature (Fig. 7).[7]

The combination of equations (14) and (17) results in a formula for the effect of time and temperature on case depths[2,35,36]

$$x = At = \frac{0.79\sqrt{Dt}}{0.24 + (C_{AT} - C_O)/(C_P - C_O)} - 0.7\frac{D}{\beta} \qquad (18)$$

assuming a defined carbon content C_{At} (limiting carbon content) at the case depth At.

For a limiting carbon content of 0.3% the case depth $At_{0.3}$ can be assessed according to Refs. 24,35

$$At_{0.3\%C} = k \cdot \sqrt{t} - D/\beta \qquad (19)$$

where k depends on temperature, carbon potential and carbon content of the core.

The carbon gradient as the major aim of carburising, determined by the case depth At and the surface carbon content C_R, can thus be established for any given temperature in the austenite phase field with the carburising time according to equation (19) and with the carbon potential. In industrial carburising the carbon

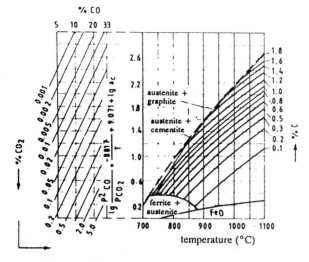

Fig. 8. Carbon potential control by means of Boudouard reaction, general relationship.[13]

potential is controlled by using[17] equation (1) (infrared analysis of the carbon dioxide content of the atmosphere) equation (2) (dew-point analysis of the water vapour content) and equation (3) (measuring of oxygen content with the oxygen probe). The slow reaction of methane (equation (4)) cannot be used for controlling. The relationship between the carbon content in austenite and the composition of the atmosphere is shown in Fig. 8[13] for equation (1) in general and in Fig. 9[18] as an example of carrier gas containing 20% carbon monoxide. Similar relationships are known for controlling the carbon potential by means of equations (2) and (3) respectively; see Refs. 13,18,36.

3 MICROSTRUCTURES OF CARBURISED COMPONENTS

In contrast to other steels with high carbon content, carburised microstructures are quenched from the homogeneous austenite phase field and are determined by the parameters, carbon gradient and quenching conditions.

The *carbon gradient* results in a gradient of martensite start temperature M_S with low M_S-temperatures in the case and high M_S-temperatures in the core. Quenching of the entire carburised component with critical cooling rate results in a microstructure distribution[37,38] with plate martensite in the case and lath martensite in the core, whose respective hardness values make up the hardness gradient. In particular with alloy steel, the high contents of soluted carbon in the austenite inevitably lead to retained austenite in the case. The diffusionless transformation of austenite into martensite takes place solely within the austenite grains; the austenite grain boundary is consequently an additional characteristic of martensite.

236 Quenching and Carburising

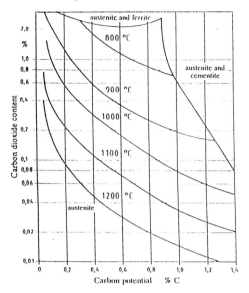

Fig. 9. Carbon potential control by means of Boudouard reaction, example for 20% carbon monoxide in the atmosphere.[18]

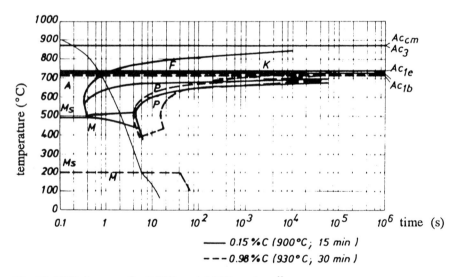

—— 0.15 %C (900°C; 15 min)
--- 0.98 %C (930°C; 30 min)

Fig. 10. TTT-diagrams for 0.15% and 0.98% carbon.[39]

The *quenching conditions*, related to a given quenchant, are determined by the dimension of the component and the hardenability of a steel. Most important are the cooling characteristics in the TTT-diagrams. Figure 10[39] contains both the

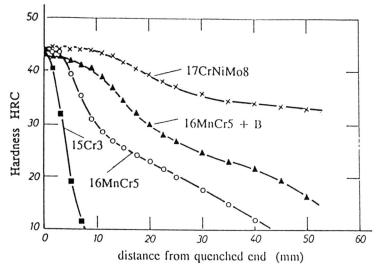

Fig. 11. Jominy test results for some carburising steels.[40]

TTT-diagram of a steel with 0.98%C (representing the case) and of a steel with 0.15%C (representing the core) and clearly shows the dominating influence of the carbon content on hardenability. In selecting carburising steels it is therefore most important to consider the alloy-dependent hardenability of the core (Fig. 11[40]).

Figure 6 shows that carbides and oxides can also develop in carburising atmospheres. The surface oxidation is particularly observed in steels alloyed with chromium.[41,42]

According to the described parameters and relationships carburised microstructures consist of plate martensite and retained austenite in the case (Fig. 12); surface oxidation is inevitable in most of the industrially used carburising atmospheres. Unetched cross-sections show that the surface oxidation affects mainly the grain boundaries (Fig. 13). In most cases the austenite grain boundary can only be shown separate from the matrix microstructure (Fig. 14). Carbides in the case (Fig. 15) are usually considered as faults. The core microstructure consists of lath martensite if quenched with critical cooling rate (Fig. 16).

According to the TTT-diagrams all kinds of microstructures can be found if the critical cooling rate is not reached.

Almost all carburised parts are tempered at about 180°C; the ensuing changes in the microstructures are not visible in the optical microscope.

4 TOUGHNESS OF THE CARBURISED MICROSTRUCTURE

The toughness of carburised composite materials is dominated by two major parameters; all the other parameters are of much less influence.

Fig. 12. Plate martensite and retained austenite in the case.

Fig. 13. Surface intergranular oxidation.

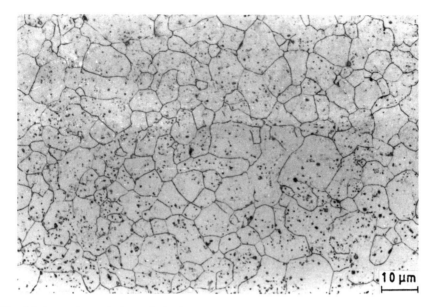

Fig. 14. Austenite grain boundaries in carburised steel.

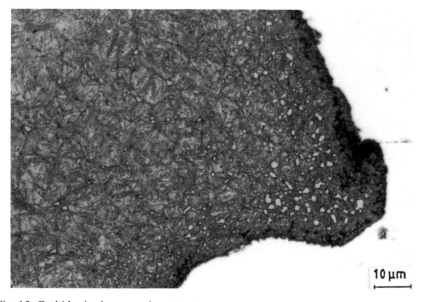

Fig. 15. Carbides in the case microstructure.

240 *Quenching and Carburising*

Fig. 16. Lath martensite in the core microstructure.

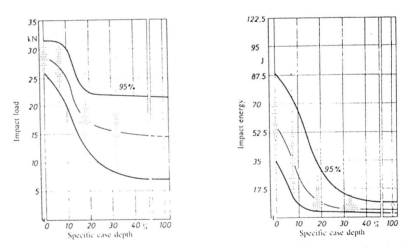

Fig. 17. Influence of case depths on impact load and absorbed energy of carburised specimens.[43]

The first dominating parameter is the distribution of the microstructures, which can be described by the specific case depth, determined as the relation of case depth in the maximum stressed cross-section to the total cross-section of the component. With an increasing amount of hard microstructure in the case the resistance of the component against impact loads decreases significantly as shown by impact loads and absorbed energy (Fig. 17[43]) measured with an instrumented impact test device.

With a specific case depth of more than 30% there is now further advantage in carburising. Comparing the toughness of carburised components, it is always necessary to refer the influence of material, design and heat treatment in particular, on the same specific case depth in order to avoid wrong conclusions. The results in Fig. 17 show also that the toughness of carburised components is described satisfactorily by the instrumented impact test, which can be carried out easily.

The second dominating parameter is the carbon content in the case and core, which determines microstructures and properties (Fig. 18[44]). Carbon contents for maximum hardenability (0.6 to 0.7%) are needed in the case and low carbon contents according to DIN 17210 and are of advantage in the core. Again it should be mentioned that only the properties of those microstructures can be compared which have equal carbon gradients.

The two results allow some important conclusions on the toughness of carburised components

- the specific case depth should be as small as possible (this will almost always conflict with the requirements of high fatigue, high contact fatigue and high wear resistance performance);
- the carbon content in the case should be close to 0.6% (a carbon content in the range of 0.6–0.7% can be easily maintained in computer-controlled carburising);
- the design should avoid single overloads, for instance by using an automative gear with a smooth load characteristic instead of a manual gear which creates much higher overloads.[45]

As components become smaller and smaller in developing new products, the imposed loads and the specific case depths become higher. Overload fracture caused by unstable cracks can usually be avoided, nevertheless stable cracks due to inevitable single overloads are dangerous and cannot be allowed because these cracks can initiate damage by fatigue, pitting and wear and shorten the in-service time of a carburised component. In particular the fatigue strength is dramatically lowered by cracks.[46]

The toughness of a carburised component with constant case depth and an already optimised carbon content can only be further improved by a higher ductility of the case microstructure. The range of results in Fig. 17 points to a possible improvement because, according to Fig. 19,[43] the higher values belong to carburising steels alloyed with nickel, whereas the lower values belong to carburising steels which are not alloyed with nickel.

5 DUCTILITY OF CARBURISED MICROSTRUCTURES

The manifold parameters on the properties of a microstructure can be better understood within the two interactions, *material* and *heat treatment*, which together make up the *microstructure* and *microstructure* and *imposed loads*, which

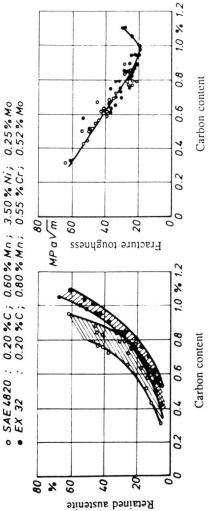

Fig. 18. Influence of carbon content on toughness.[44]

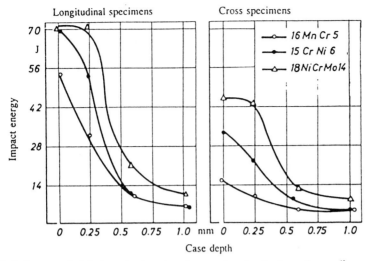

Fig. 19. Influence of nickel content on toughness of carburised specimens.[43]

in turn define *properties* and *performance*. The topic of this presentation will be those parameters and interactions which are important to the ductility.

The *material*, i.e. the carburising steels, involves mainly the alloying elements carbon, as discussed before, manganese, chromium, nickel, boron and aluminium (affects the grain size); the impure elements phosphorous and nitrogen and inclusions (with sulphur and oxygen). The effects of alloying nickel and those of phosphorous will be discussed.

Heat treatment can be described by the parameters of carburising (carbon diffusion and quenching); it sometimes has side effects like surface intergranular oxidation[41,42] and charging with hydrogen.[47] The results to be presented are concerned with gas carburising and direct quenched or reheated and quenched (double quenched) microstructures.

Imposed loads critical to ductility are mainly bending forces with slow and impact loading. Slow bend and impact tests are therefore suitable to characterise ductility. Often the bending stress which initiates the first crack in the case (bending–crack–stress) is used advantageously to describe ductility, in particular when there is a direct interaction between bending–crack–stress and fatigue strength.[48]

Distribution, development and the main properties of the carburised *microstructure* depend on the carbon content; as discussed earlier, it is therefore easy to understand why the ductility, too, depends directly on the carbon content (Fig. 20[49]).

Bending–crack–stresses from slow bend tests (Fig. 20[49]) and impact loads from impact tests (Fig. 21[49]) of microstructures with equal carbon gradients (Fig. 1) and hardness gradients (Fig. 2) increase with increasing nickel content.[49] Both tempered and double quenched microstructures are more ductile than the direct

244 *Quenching and Carburising*

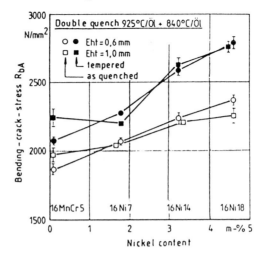

Fig. 20. Influence of retained austenite on bending strength.[49]

quenched and the as-quenched structures. The influence of the case depth is clearly to be seen. The influence of nickel on the bending–crack–stress is more distinctive than its influence on the impact load; it has an effect only when more than 2% nickel are alloyed.[42,49]

The beneficial effect of nickel–molybdenum has a similar effect,[50,51]; chromium tends to be disadvantageous[43] – this can be explained for several reasons.

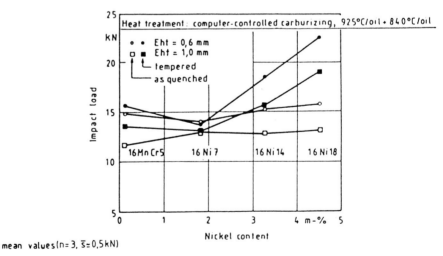

Fig. 21. Influence of nickel content on bending–crack–stress.[49]

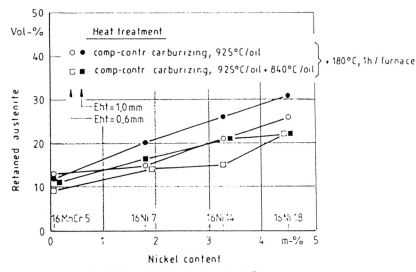

Fig. 22. Influence of nickel content on impact load.[49]

Obviously the amount of retained austenite increases with higher nickel contents (Fig. 22[49]). Retained austenite transforms under bending stress into martensite, the increasing volume creates additional compressive stresses, which contribute to a higher bending–crack–stress.[52,53] This process is time-dependent and probably is a reason for the different influence of nickel on bending–crack–stress and impact load. Furthermore nickel has a delaying effect on carbide formation during quenching from carburising or austenitisation temperature[54] and improves, in particular, in tempered, less distorted microstructures, the substructure of martensite to a more homogeneous dislocation movement.[55] With increasing nickel content more phosphorous was segregated on grain boundaries but did not lead to temper embrittlement and thus did not influence the presented results. In accordance with the discussed results and reasons the fracture mode shifted with increasing nickel content from a mainly intergranular fracture (Fig. 23) in structures with low nickel content to a transgranular fracture (Fig. 24) in structures with more than 2% nickel.

The macro residual stresses (Fig. 25[49]) of the examined carburised microstructures are found within a small scatter band and are independent of the nickel content; the residual stresses have therefore no influence on ductility, which improves with increasing nickel content.

Results from bending tests (Fig. 26[56]) on carburised specimens with different case depths (residual stresses according to Fig. 25) and on through hardened specimens (with uniform tensile residual stresses of about 100 N mm^{-2}) show that bending-crack-stresses are independent of the related case depth and the macro residual stresses. These results prove that the bending-crack-stress describes only the behaviour of the case microstructure, which consists of plate martensite and

246 *Quenching and Carburising*

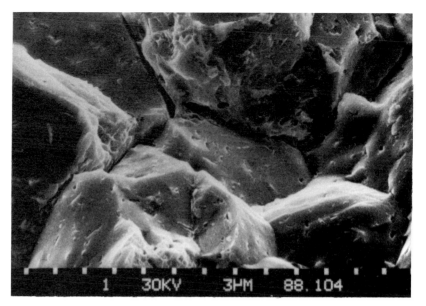

Fig. 23. Amount of retained austenite in nickel alloyed cases.

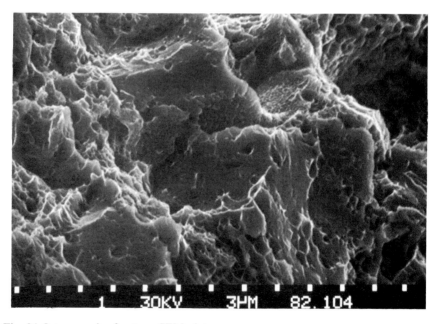

Fig. 24. Intergranular fracture, SEM picture.

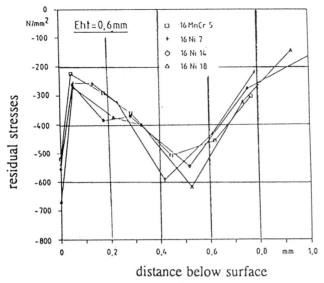

Fig. 25. Transgranular fracture, SEM picture.[49]

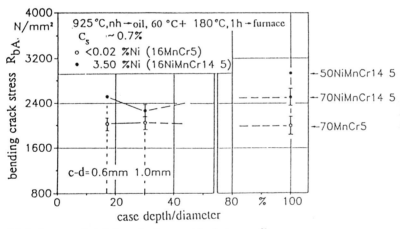

Fig. 26. Influence of nickel content on residual stresses.[56]

retained austenite. The beneficial effect of increased nickel contents (and of low carbon contents) on ductility is shown clearly in Fig. 26.

6 CONCLUSIONS OF TOUGHNESS AND DUCTILITY

The toughness of a carburised component is defined by its specific case depth, which describes the amount of the hard case in the component, and by the carbon content in the case and core. Small specific case depths and a low carbon content, within the limits of carburising, of course, result in a good toughness performance. The case depth, in turn, depends largely on the main properties of the carburised components – fatigue strength, contact fatigue strength and wear resistance; usually there has to be a compromise to meet the different requirements.

Given a fixed case depth and optimal carbon content, the toughness is additionally defined by the ductility of the case and can be significantly improved by alloying more than 2% nickel without any loss of hardness in the case(!). Within the given context of these interrelations tempered and double quenched microstructures are superior to non-tempered and direct quenched microstructures. Bending–crack–stresses from slow bend tests and impact loads from impact tests are found to present the ductility sufficiently, the bending–crack–stress correlates directly with the fatigue strength.

REFERENCES

1. B. Thoden and J. Grosch: *Neue Hütte*, 1989, **34**, 96–98.
2. J. Wünning: *Häterei Techn. Mitt.*, 1968, **23**, 101–109.
3. F.E. Harris: *Metals Progress, Jan 1945*, **77**, 84–89.
4. R.W. Gurry: *Trans. AIME, April 1950*, **188**, 671–687.
5. Th. Schmidt: *Härterei Techn. Mitt. 1952*, Sonderheft Gasaufkohlung 11–30.
6. F. Neumann and B. Person: *Härterei Techn. Mitt. 1968*, **23**, 296–310.
7. F. Neumann and U. Wyss: *Härterei Techn. Mitt. 1970*, **25**, 253–266.
8. F.J. Harvey: *Metall. Trans. 1978*, **9A**, 1507–1513.
9. U. Wyss: *Härterei Techn. Mitt. 1962*, **17**, 160–171.
10. H.V. Speck, H.J. Grabke and E.M. Müller: *Härterei Techn. Mitt. 1985*, **40**, 92–103.
11. F. Neumann: *Härterei Techn. Mitt. 1978*, **33**, 192–201.
12. O. Schaaber and R. Fischer: *Industrieblatt, 1956*, **2**, 89–96.
13. U. Wyss: *Härterei Techn. Mitt.*, 1990, **45**, 44–56.
14. E. Schürmann, Th. Schmidt and H. Wagner: *Gießerei, Beiheft 1964*, **16**, 91–95.
15. S. Gunnarson: *Harterei Techn. Mitt. 1967*, **22**, 292–295.
16. K.H. Sauer, M. Lucas and H.J. Grabke: *Härterei Techn. Mitt. 1988*, **43**, 45–53.
17. F. Neumann: *Härterei Techn. Mitt. 1989*, **44**, 262–269.
18. B. Edenhofer: *Härterei Techn. Mitt. 1990*, **45**, 119–128.
19. A. Fick: *Poggendorf Annalen, 1885*, **94**, 59–64.
20. C. Wells and R.F. Mehl: *Metals Technology (NY), Techn. Publ. 1180, Aug 1940*.
21. A. Slattenscheck: *Härterei Techn. Mitt. 1942*, **1**, 85–135.
22. J.I. Goldstein and A.E. Moren: *Metall. Trans.*, 1978, **9A**, 1515–1525.
23. J. Pavlossoglou and H. Burkhard: *Härterei Techn. Mitt. 1976*, **31**, 209–213.
24. H.U. Fritsch and H.W. Bergmann: *Härterei Techn. Mitt. 1986*, **41**, 14–20.
25. A.R. Marder, S.M. Perpetua, J.A. Kowalik and E.T. Stephenson: *Metall. Trans. 1985*, **16A**, 1160–1163.
26. C.A. Stickels: *Heat Treatment and Surface Engineering*, ed. G. Krauss, *ASM*, 1988. S.99–102.
27. T. Reti and M. Cseh: *Härterei Techn. Mitt. 1987*, **42**, 139–146.
28. J. Wünning: *Härterei Techn. Mitt. 1985*, **40**, 104–105.
29. R. Hoffmann: *Härterei Techn. Mitt. 1979*, **34**, 130–137.

30. J. Wünning: *Z. wirtsch. Fertigung, 1982,* **77,** 424–426.
31. V. Schüler, B. Huchtemann and E. Wulfmeier: *Härterei Techn. Mitt. 1990,* **45,** 57–65.
32. J. Grosch: *Härterei Techn. Mitt. 1981,* **36,** 262–269.
33. J. Pavlossoglou: *Härterei Techn. Mitt. 1976,* **31,** 252–256.
34. H.J. Grabke: *Härterei Techn. Mitt. 1990,* **45,** 110–118.
35. F.E. Harris: *Metal Progress 1943,* August, 265–272.
36. *Metals Handbook*, 9th edn, Vol 4, *Heat Treating*, 135–175.
37. G. Thomas: *Metall. Trans. 1971,* **2,** 2373–2385.
38. G. Krauss: *Härterei Techn. Mitt. 1986,* **41,** 56–60.
39. A. Rose and H. Hougardy: *Atlas zur Wärmebehandlung von Stählen*, Vol. 2, Verlag Stahleisen, Düsseldorf, 1972.
40. H. Dietrich and W. Schmidt: *Thyssen Techn. Berichte, 1984,* **10,** 105–132.
41. R. Chatterjee: *Z. wirtsch. Fertigung, 1976,* **71,** 367–372.
42. R. Chatterjee-Fischer: *Metall. Trans., 1978,* **9A,** 1553–1560.
43. D. Wicke and J. Grosch: *Härterei Techn. Mitt. 1977,* **32,** 223–234.
44. Y.E. Smith and D.E. Diesburg: *Metal Progress, May 1979,* 68–73.
45. H. Brugger: *Schweizer Archiv angew. Wiss. 1970,* **36,** 1–11.
46. H. Weigand and G. Tolasch: *Härterei Techn. Mitt. 1967,* **22,** 330–338.
47. H. Streng, C. Razim and J. Grosch: *Härterei Techn. Mitt. 1987,* **42,** 245–259.
48. J.L. Pacheo and G. Krauss: *Härterei Techn. Mitt. 1990,* **45,** 77–83.
49. B. Thoden and J. Grosch: *Härterei Techn. Mitt. 1990,* **45,** 7–15.
50. M. Bacher, O. Vöhringer and E. Macherauch: *Härterei Techn. Mitt. 1990,* **45,** 16–29.
51. D.E. Diesburg and Y.E. Smith: *Metal Progress, June 1979,* 35–39.
52. H. Brandis: *et al. Härterei Techn. Mitt. 1983,* **38,** 63–67.
53. M.A. Zaccone, J.B. Kelly and G. Krauss: *Heat Treatment*, 87, The Institute for Metals. London 1988, 93–101.
54. H.K. Obermeyer and G. Krauss: *J. Heat Treating 1980,* **1,** 31–39.
55. H. Kwon and C.H. Kim: *Metall. Trans. 1986,* **17A,** 1173–1178.
56. O. Schwarz, B. Thoden and J. Grosch: *Härterei Techn. Mitt., 1991,* **5.**

15

Martempering and Austempering of Steel and Cast Iron

GEORG WAHL

Leybold-Durferrit GmbH, Rodenbacher Chaussee 4, D-6450 Hanau-Wolfgang, Germany

ABSTRACT

Martempering and austempering of steel and cast iron give savings in heat treating costs, improved component properties, less distortion and greater economy. The limits of the process are determined by the composition of the material and the transformation behaviour.

The austempering of cast iron is beginning to be used in the transmission field, in mechanical and apparatus engineering and in the automative industry.

1 INTRODUCTION

Essential for a good quality heat treatment is a controlled and uniform cooling of the whole component. The cooling speed necessary to form martensite or bainite and avoid premature and unacceptable transformations of the structure is determined by the composition of the steel and cast iron used. Oil or martempering salts are usually employed for cooling alloyed steels.

Cooling at a temperature below the martensite point, Ms, causes heavy internal stressing and distortion. Salt baths do not have these disadvantages. When cooling and holding at 180–200°C while martempering a temperature equalisation takes place first of all over the whole cross-section of the workpiece. During further slow cooling in air to room temperature a low stress martensite transformation of the surface layer takes place.

It should be borne in mind that the surface and the core of case hardening steels transform at different temperatures due to the different carbon contents.

2 MARTEMPERING IN SALT BATHS

In salt quenching the heat transfer mainly takes place by convection. A boiling period, such as experienced with quenching oils, does not occur. Therefore the cooling characteristics of oil are quite different to those of salt martemper baths.

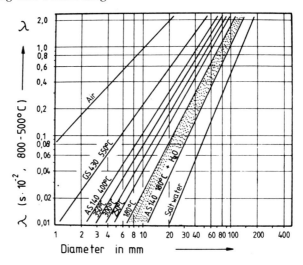

Fig. 1. Cooling parameter for AS 140 martempering salt.

The development of martempering oils also makes it possible to use higher temperatures but, however, there is a visible trend away from oil to salt martemper baths in the treatment of components which tend to distort. A further argument could be that it is easier to control the disposal and treat the effluents.

The cooling intensity of salt martemper baths can be defined by the factor λ. As demonstrated in Fig. 1, the cooling effect is influenced by the bath temperature and water content. In order to use the benefits of marquenching on unalloyed steels it is the usual practice to improve the cooling effect by adding 0.5–1.0 weight % water. This measure considerably broadens any applications for salt martemper baths. Higher water contents do not improve the cooling effect further and are not recommended for safety reasons.[1]

Figure 1 shows the influence of the water content at a bath temperature of 180°C on the cooling intensity.

An automatic probe has been developed for regulating the cooling intensity of salt martemper baths by adding water. It replaces the traditional empirical method of adding water. By continuously monitoring and controlling the salt martemper bath, reproducible and defined cooling conditions to match individual technical needs are possible within the temperature range 180–250°C.

The measuring principle is based on the simulated cooling of a workpiece. Basically the controlling device consists of a measuring probe, electronic control system and water feeder.[2]

The sensor (Fig. 2) is put into the martemper bath and heated up to 200°C above bath temperature by the built-in heating. Due to the difference in temperature the probe gives off a certain amount of heat to the salt bath, which is cooler. The amount of heat given off depends on the cooling effect of the martemper bath. Influences due to other salts being brought into the bath are also taken into

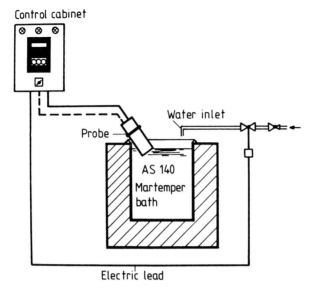

Fig. 2. Salt martemper bath – control unit.

account without difficulty. The amount of heat emitted – which is time related – enables a measured variable to be derived which is very suitable for regulating the water content and thus the state of an AS 140 martemper bath.

Approximately 20 minutes after being switched on the unit is ready for use. A measuring cycle runs through about 20 times/hour and the actual state is shown on the digital display. The desired value is set and a certain amount of water added to the martemper bath via a controlled magnetic valve when the actual value differs from the set value. This control system has proved to be reliable under various working conditions.

3 SALT MARTEMPER BATHS COMBINED WITH CONTROLLED ATMOSPHERE PLANTS

In the heat treatment of parts which tend to distort, the benefits of salt martemper baths are used with controlled atmosphere furnaces. On choosing the appropriate quenching medium, all factors, for example the necessity of employing mandrel hardening, post machining and rejects or omission of meshing times etc., must be taken into account. A practical example of combining controlled atmosphere with a salt bath is shown by the schematic diagram in Fig. 3.[3]

In order to compare hardening in oil with hardening in a salt martemper bath the plant was equipped with two different quenching facilities and operated alternately – oil hardening/salt martempering – throughout the test period. Furnace-related or process-related fluctuations were thereby excluded. The methods of measuring used in the comparison are described in Fig. 4.

254 *Quenching and Carburising*

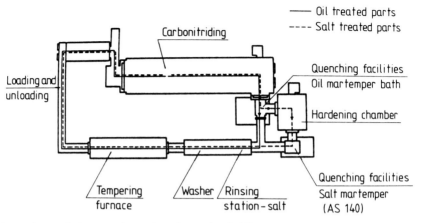

Fig. 3. Controlled atmosphere plant for carbonitriding.

Two different sized ring gears made from SAE 4027-H were used for the test. Cooling in the oil martemper bath was done at 160°C and in the salt martemper bath at 210°C. The differences in temperature were due to the fact that the martemper oils available for low alloyed steels did not produce constant hardening results at temperatures above 160°C. Fluctuations in core hardness and case depths were clear evidence of this. Furthermore, it was noticed that the martemper oils had a shorter lifetime which, among other things, resulted in non-uniform hardening results.

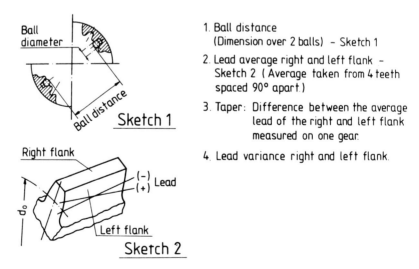

Fig. 4. Processes for measuring the dimensional changes during hardening.

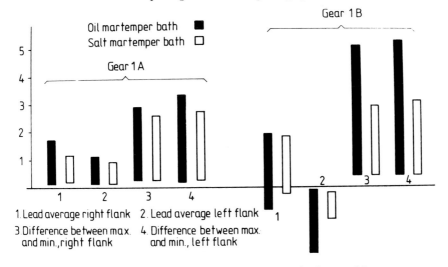

1. Lead average right flank 2. Lead average left flank
3. Difference between max. 4. Difference between max.
   and min., right flank   and min., left flank

Fig. 5. Changes in dimension during oil martempering and salt quenching.

A comparison of the changes in dimension which take place during hardening, measured by the distance between two balls mounted in a crown gear, showed distinct advantages in favour of salt quenching. This tendency is also confirmed in Fig. 5 by the tooth lead.

These good results encouraged the user to convert other carbonitriding plants to salt martempering, which too are producing high quality heat treating results. Hardnesses and dimensions are within the specified tolerances. The difference in viscosity between martempering oils and salt martemper baths causes differences in drag out at martempering temperature. Greater losses can be expected with salt martempering baths than with oil.[4]

A real help in minimising the drag-out losses are suitably designed jigs, arrangement of the charges and adequate drip off times over the martemper bath.

Particular attention should also be paid to the cleaning of the components and in particular the jigs. Only careful cleaning will ensure that no trouble occurs in the gas atmosphere after salt martempering.

4 AUSTEMPERING

With this process the desired heat treated state is not obtained by tempering a martensitic structure. In actual fact it is a treatment by which, after through heating to hardening temperature, the existing austenite is transformed isothermally into bainite.[5-8]

This is done by putting the parts, while still at hardening temperature, into a salt bath at 250–450°C. In critical cases the Duplex Quench process[9,10] can be used to avoid preliminary transformation. First of all the parts are cooled for a short time

256 *Quenching and Carburising*

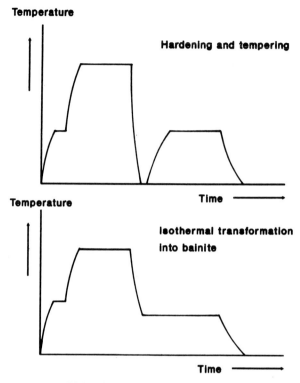

Fig. 6. Treating sequence – H & T/austempering.

in a low-temperature salt bath. A second salt bath takes care of the isothermal transformation into bainite.

The transformation of the austenite into bainite does not take place suddenly within this temperature range, as in the formation of martensite. Bainite formation is linked with time-related precipitations. Therefore workpieces stay isothermic for periods ranging from some minutes to a few hours at the same temperature in the austempering bath. On completion of the transformation, further cooling to room temperature can take place. In practice this is usually done in water to rinse off the salt adhering to the parts at the same time. By using evaporators the salt regained from the water tanks can be returned to the austempering bath.

Figure 6 compares austempering with normal hardening and tempering. Holding time and austempering temperature during transformation into bainite are of considerable importance for a faultless austempering. The temperature at which bainite is formed mainly influences the hardness and strength of the material being treated.

Figure 7 demonstrates the transformation in relation to the composition of the steel, and shows the isothermal transformation and continuous cooling of unalloyed plain carbon steel. The necessary cooling conditions can be taken from the lower diagram for continuous cooling. To prevent pearlite formation the

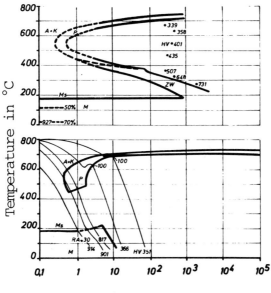

Fig. 7. TTT diagram of plain carbon steel, 1%C.

temperature range between 800 and 500°C must be passed through within half a second. If the continuous cooling takes place slower than this the hardness obtained will drop and the pearlite will increase. Martensite (Ms) begins to form in this steel at approximately 180°C.

The upper diagram of Fig. 7 shows an isothermal transformation. Such a transformation into pearlite is not possible because there is no cooling agent available for this unalloyed steel which permits rapid cooling from 800 to 500–600°C without preliminary transformation occurring. Isothermal transformation into bainite takes place within a temperature range of 200–350°C.

Figure 8 shows the TTT diagram of ball bearing steel AISI L3. The influence of chrome as an alloying element can be clearly seen. Due to the low chrome content of 1.5%, the transformations take much longer and at higher temperatures. During continuous cooling, pearlite begins to form after approximately 60 seconds at 600°C. Furthermore this steel has a distinct bainite range. Due to the change in composition, the temperature at which martensite (Ms) begins to form is raised to approximately 250°C compared with the 180°C of the 1% carbon steel.

As TTT diagrams show, the holding time for the isothermal transformation into bainite is determined above all by the composition of the material. Steel parts used in bainite hardening usually contain 0.3–0.7%C; in special cases even more. Unalloyed steels are treated, also those materials alloyed with chrome and/or molybdenum and nickel.

The alloying elements influence the heat treating parameter and the component properties. As a general guide it can be said that the higher the alloy content of

258 Quenching and Carburising

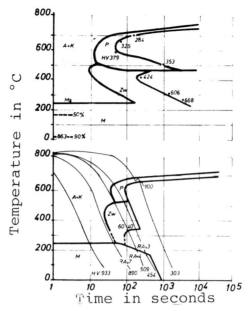

Fig. 8. TTT diagram of steel AISI L3.

thick cross-sectioned components the longer the transformation into bainite will take. Longer treating times mean, however, higher costs. Therefore, due to the long transformation times, high alloyed materials are rarely austempered.

4.1 INFLUENCE OF AUSTEMPERING ON THE COMPONENT PROPERTIES

In the broad range of literature available on this subject, austempering is described as having benefits with regard to economy and component properties over normal hardening and tempering. Compared with normal hardening and tempering – based on equal strength or hardness – the heat treating time is shorter and toughness values such as elongation, necking, bending behaviour and notch impact strength and fatigue strength are improved.

Figure 9 shows the notch impact strength of material SAE 6150 over the hardness in relation to the heat treatment carried out. The range of 42–55 HRC shows the distinct superiority of austempering over hardening and tempering. Beyond these limits, in other words at higher or lower hardness values, the result is influenced by unacceptable pretransformations. Therefore it is important in practice, by means of the TTT diagrams and possible heat treating conditions, to know and specify the limits. A reliable aid when using the TTT diagrams are the values which give an indication of the cooling obtainable with the cooling medium used, taking into account the dimensions of the component.

The toughness properties determined from samples made from SAE1078 with a

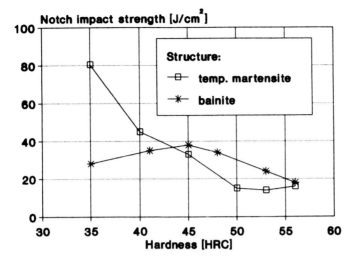

Fig. 9. Notch impact strength of SAE 6150.

strength of approximately 1800 N/mm² (comparable with approximately 50 HRC) are shown in Fig. 10. In spite of the relatively high hardness, necking and notch impact strength are distinctly above the values of the normal hardened and tempered components. Also very interesting is the maximum bending angle until cracking occurs – it is three times greater after austempering than after normal hardening and tempering.

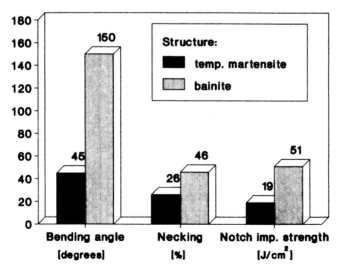

Fig. 10. Properties of steel SAE 1078.

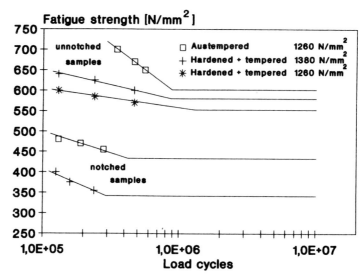

Fig. 11. Fatigue strength of 30 Si Mn Cr 4.

Essential for the high toughness of a steel is the complete transformation into bainite. If the holding time in the austempering bath is too short, this will result in mixed structures, e.g. bainite and martensite with lower toughness. The same also applies of course to unacceptable pretransformation due to insufficient cooling speed.

The dynamic loadability of components is also favourably influenced by austempering. References have been made to increases in the time-related fatigue strength and fatigue strength. Presumably this is attributable to the low notch sensitivity of the bainite structure (Fig. 11).

4.2 APPLICATIONS FOR AUSTEMPERING

This process has proved to be most successful, particularly in the manufacture of small parts in high-volume production. Table 1 gives an impression of the range of unalloyed and low alloyed steels treated, with wall thicknesses preferably less than 5 mm but in special cases up to approximately 20 mm thick.

Some of the parts listed in Table 1 are shown in Figs. 12 and 13. The hard metal tipped rock drills are austempered to improve their toughness. A further benefit is that excess stresses, which could trigger off cracks in the welded joint or in the hard metal insert, are avoided.

4.3 AUSTEMPERED NODULAR CAST IRON

Austempered ductile iron has become increasingly popular in recent years in the manufacture of components. In the middle of the 1960s investigations were started

Table 1.

Workpiece	Material	Thickness (mm)	Average HRC
Spring washers, various sizes	50CrV4	2–5	45–50
Spring rings, various sizes	Ck60	0.5–2	42–50
Seeger circlips	Ck67	1–3	44–48
Adapter sleeves	C60	1–2	45–50
Notched pins	C90	2–5	55–58
Fasteners	Ck67	0.5–2	46–48
Compensating springs	Ck75	1	46–48
Belt links	67Si7, Ck75	1–1.5	46–50
Chain links	Ck67–75	1–3	44–48
Office machinery parts	Ck55–80	0.5–2	40–50
Type bars	67SiCr5, C55	1	40–50
Concrete nails	41Cr4, 50CrV4	5	54–57
Crown nuts	Ck45	6	30–35
Mower plates	Ck67	2–3	52–55
Chain saw parts	Ck67	1–2	50–56
Hinges and fittings	Ck67	1–3	42–48
Hard metal tipped drills	34CrNiMo6	20	42–48
	50CrV4	15	42–48

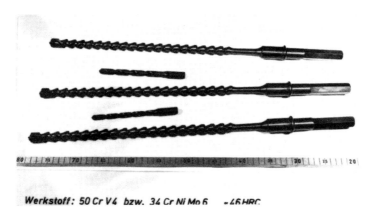

Fig. 12. Austempering of rock drills.

Fig. 13. Various small parts – austempered.

in the American automobile industry to substitute case hardened steels used for transmissions for austempered ductile iron. In a publication on the austempering of ductile iron John Dodd reports the use of this material for rear axle gears and pinions in passenger cars instead of case hardened steel parts. There are a number of publications[11,12] on the use of austempered ductile iron for making transmissions. Apart from gear wheels, reference is also made to the use of other components such as bearing shells, crankshafts, wheel hubs and railway axle boxes with the following benefits:

- minimising of mechanical processing by casting the components;
- less noise due to better damping;
- savings in costs due to easier heat treatment;
- good processing properties;
- savings in weight.

Cast iron is also best austempered within the temperature range 250–450°C. Depending on their composition, cast iron qualities with wall thicknesses of up to approximately 50 mm are treated. Table 2 gives details of the approximate content of the alloying element required for the through hardening of nodular cast iron containing 3.3% C, 2.4% Si and 0.32% Mn.[13-16]

4.4 PLANTS FOR AUSTEMPERING

There are a number of different plant systems – continuous and non-continuous – available nowadays for carrying out heat treating. This ensures that there are enough possibilities of matching heat treatment to material, component shape and

Table 2. Approximate content of alloying element for the hardening through of nodular cast iron containing 3.3% C, 2.4% Si, 0.32% Mn

Wall thickness (mm)	Alloying content in % cooling medium salt bath
10	Unalloyed
25	0.03 Mo
37	0.05 Mo
	0.35 Mo + 1 Cu
50	0.05 Mo + 1 Cu

the required throughput. Most common are plants in which austenitising is done in protective gas and the isothermal transformation into bainite in a salt martemper bath. In many cases salt bath plants alone have proved themselves. Figure 14 shows a schematic diagram of an automated salt bath plant for the austempering of guide tracks of wood processing machines. This plant permits a throughput of 2500 kg/hour.

Salt melts have also been used most successfully for austempering wire coils. For some years now a German steel producer has been operating a salt bath in which 45 t/hour of wire coils each weighing between 400 and 1200 kg can be treated.[17]

For treating long slim parts multi-purpose pit-type furnaces are available in

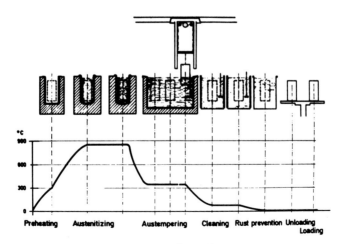

Fig. 14. Treating sequence – automated salt bath plant.

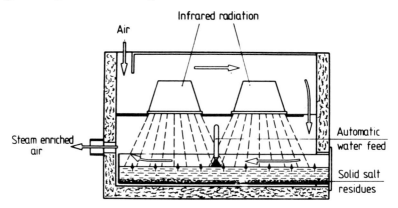

Fig. 15. Infrared evaporator.

which, after austenitising in protective gas, the isothermal transformation into bainite can be carried out in salt martemper baths.[18-20]

These continuous plants available include shaker hearth furnaces, rotary furnaces and conveyor-belt type furnaces. An effluent-free and in many cases also a waste-salt free heat treatment has already been achieved by combining protective gas with marquenching. Evaporator systems for processing the salt-bearing rinsing waters, with the possibility of reusing the regained salt, are available.

Infrared evaporators, which make traditional detoxifying plants superfluous, have proved very successful. Rinse waters enriched with martemper salts are vaporised in the evaporator by infrared rays at temperatures below boiling point within a very short time. This treatment leaves behind solid salts, which can be reused. The evaporator can be gas-fired or electrically heated, is of compact design, is easy to handle, requires little space and works fully automatically (Fig. 15). An outlet valve facilitates easy removal of the regained salt.

REFERENCES

1. K. Winterer: 'Die Erhöhung der Abkühlwirkung von AS-Warmbädern und ihre Messungen', *Durferrit-Hausmitteilungen, 1957*, **30**, 21–24.
2. W. Huth: 'Erfahrungen mit einem Gerät zur kontinuierlichen Messung der Abkühlintensität von Warmbädern', *Vortrag Härterei Kolloquium, Wiesbaden, 1986*.
3. H. Schlösser: 'Einsatz von Abschrecksalz AS 140 und Hochtemperaturabschrecköl in einer Getriebehärterei', *Durferrit-Hausmitteilungen, 1978*, **43**, 18–21.
4. H. Beitz and U. Rettberg: Ölwarmbäder für moderne Ofenanlagen 'neue' Fachberichte, 13. Jahrgang, 642–651.
5. K. Falkenmayer: Praktische Anwendung von isothermischen ZTU-Schaubildern ZwF 75, 1980, **9**, 434–438.
6. J. Motz: 'Die Kohlenstoffauflösung im Austenit von Gußeisenlegierungen und ihre Bedeutung für die Wärmebehandlung', *Giesserei, 10/57*, **18**, 943–953.

7. W. Mannes, K. Hornung and H. Rettig: 'Erprobung von Zahnrädern aus unlegiertem bainitischem Gußeisen mit Kugelgraphit', *Konstruieren + Gießen 1985*, **10** (4), 19–29.
8. H. Mühlberger: 'GGG 100 B/A – Gießen statt Schmieden', *Fachberichte für Metallbearbeitung, 1985*, **62** (11/12), 658–669.
9. C. Skidmore: 'Salt Bath Quenching – a Review', *Heat Treatment of Metals 1986*, **2**, 34–38.
10. W. Hauke: 'Isothermisches Umwandeln von Gußeisen mit Kugelgraphit in der Bainitstufe', *Härtereitechn. Mitteilungen, 1983*, **37**, 72–77.
11. M. Johnsson: 'Austenitisch-bainitisches Gußeisen mit Kugelgraphit als Konstruktionswerkstoff im Getriebebau', *Antriebstechnik, 1976*, **15** (11), 593–600.
12. J.M. Motz: 'Bainitisch-austenitisches Gußeisen mit Kugelgraphit – hochfest und verschleißbeständig', *Konstruieren + Gießen, 1985* **10** (2), 4–11.
13. J. Dodd: 'Zwischenstufenvergütung von Gußeisen mit Kugelgraphit', *Gießerei 1978*, **65**, 73–80.
14. W.W. Cias: Unveröffntl. Forsch. Ber. der Climax Molybdenum Company of Michigan.
15. J.W. Boyes and N. Carter: *British Foundary, 1966*, **59** (9), 379–86.
16. W. Scholz and M. Semchyshen: *Mod. Cast., 1968*, **53** (1), 65–72.
17. K.-J. Kremer, K. Neuhaus, E. Sikora and M. Wirth: 'Salzbadbehandlung von Walzdraht', *Stahl und Eisen 1990* **110** (6), 51–56.
18. K. Heuertz: 'Mehrzweck-Schachtofen-Automat mit gasdichter beheizter Umsetzvorrichtung', *Härterei-Techn. Mitteilungen, 1987*, **42** (3), 169–173.
19. F.-W. Eysell: 'Die Zwischenstufenvergütung, und ihre betriebliche Anwendung', *TZ für praktische Metallverarbeitung, 1972*, **66** (3), 94–99.
20. G. Wahl: 'Entwicklung und Anwendung von Salzbädern für die Wärmebehandlung von Einsatzstählen', *AWT-Tagung EINSATZHÄRTEN in Darmstadt, April 1989*.

16
Property Prediction of Quenched and Case Hardened Steels Using a PC

T. RÉTI

Department of Material Science, Bánki Donát Polytechnic, H-1081 Budapest, Népszinház u.8, Hungary

M. GERGELY

SACIT Steel Advisory Centre, H-1116 Budapest, Fehérvári ut 130, Hungary

C.C. SZILVASSY

Faculty of Engineering, Queensland University of Technology, Brisbane, Qld 4000, Australia.

ABSTRACT

Two computerised methods which can be used for simulating the metallurgical processes occurring during heat treatment are outlined. The first program, PREDQUENCH, permits the prediction of the progress of transformations, of the microstructure and of the mechanical properties after quenching and tempering. This computational algorithm calculates the transformation process of austenite during continuous cooling. The second program, PREDCARB, is devoted to predict the process parameters of gas carburising and calculates the carbon and hardness profile developed in the workpiece. The application of the computer programs which can be run on IBM-compatible personal computers is demonstrated by practical examples.

1 INTRODUCTION

In the last two decades there have been many computerised data banks created in different countries that tried to help the metallurgists and material engineers with bibliographic and factual databases. To a certain extent, independent of this activity, but in the same time period, the computerised property predictors were starting to be developed. The computer techniques made it possible to re-evaluate the conventional formulas, to expand their application fields and to develop more complex and accurate prediction methods.

268 *Quenching and Carburising*

In this paper two computerised methods which can be used for simulating the metallurgical processes occurring during heat treatment are outlined.

The first program, PREDQUENCH, permits the prediction of the progress of transformations, of the microstructure and of the mechanical properties after quenching and tempering. This computational algorithm predicts the characteristics of the TTT diagram, and calculates the transformation process of austenite during continuous cooling. The second program, PREDCARB, is devoted to predict the process parameters of gas carburising and calculates the carbon and hardness profiles developed in the workpiece. The possibilities of application of the two computer programs are demonstrated in examples.

2 PREDICTION OF PROPERTIES AFTER QUENCHING AND TEMPERING

In what follows, a property-prediction software system (PPS) used for simulating the metallurgical processes occurring during quenching and tempering and predicting the final mechanical properties is described. The PPS consists of several program modules, which form a logical chain for property prediction. The programs of the PPS are based on a phenomenological model of kinetic of transformations taking place in non-isothermal conditions.[1,2] The program permits the prediction of the progress of transformations, of the microstructure, and of the mechanical properties as a function of time and of position in the cross-section of the heat-treated workpiece.

2.1 MODEL DESCRIPTION

The block diagram of the PPS is shown in Fig. 1. The upper part of the diagram refers to the processes occurring during austenitisation and quenching, the lower part refers to tempering processes. The 'black boxes' of the quenching and tempering calculation unit are within the frames shown with dotted lines. The programs that form the system are numbered 1 to 12.

The input data are seen on the left side of the block diagram. They are as follows:

- *Chemical composition* of the part to be hardened.
- *Initial state of microstructure* (annealed, normalised, quenched and tempered)
- *Geometry:* shape and size of the part (round bar or plate) characterised by its diameter or thickness, and distance from the cooled surface to the point where the microstructure and properties are to be predicted.
- *Heating conditions:* heating medium, austenitising temperature and the total time spent by the workpiece in the austenitising furnace.
- *Quenching conditions:* the H_R value (relative heat transfer coefficient) specifying the cooling severity of the quenchant, the temperature of quenchant, the surface heat flux density ϕ as a function of the surface temperature of the part.
- *Tempering conditions:* the tempering temperature and the duration of tempering.

COMPOSITION:		C:41CR4
C % =	0.360 - 0.470	0.410
Si% =	0.070 - 0.430	0.250
Mn% =	0.560 - 0.940	0.750
P % =	0.001 - 0.040	0.018
S % =	0.001 - 0.040	0.018
Cr% =	0.850 - 1.250	1.050
Ni% =	0.080 - 0.400	0.130
Mo% =	0.020 - 0.120	0.042
V % =	0.001 - 0.080	0.040
Cu% =	0.100 - 0.300	0.150

GEOMETRY: bar-1, plate-2, Jominy-3
bar diameter, mm   100.0
distance from surface, mm
                                       25.0

INITIAL STATE:
annealed (1), normalized (2)   2
AUSTENITIZATION:
temperature, °C   850
time, min   60

QUENCHING AND TEMPERING:	
cooling int. 1/inch	1.00
quenchant temp. °C	20
tempering temp. °C	500
tempering time, min	120
Continue Y-CR?	

A1 = 739   A3 = 787
Bs = 570   Ms = 330
                    Cp(sec)= 57.6
AUSTENITE GRAIN SIZE   0.029520 mm,
corresponding to   8 ASTM No.
FERRITE+PEARLITE= 21.6 %
BAINITE= 30.3 % MARTENSITE= 46.6 %

HARDNESS (QUENCHED)	43	HRC
HARDNESS (TEMPERED)	33	HRC
TENSILE STRENGTH	1021	MPa
YIELD STRENGTH	850	MPa
ELONGATION (A5)	14	%
REDUCTION IN AREA	50	%
IMPACT ENERGY (KU)	52	J

Fig. 2(a). Printout of input data and predicted properties generated by PREDQUENCH program.

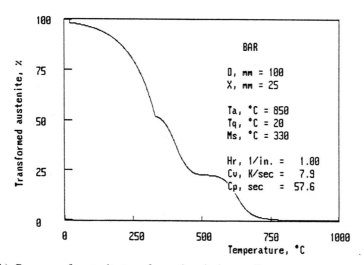

Fig. 2(b). Progress of austenite transformation during cooling.

sufficient reliable correlation, between calculated hardenability and hardenability determined experimentally, and to be used as a basis for material selection.

Example 3 (generation of transformation diagrams). By running the program with systematic changes in the input data for the distance from the cooled surface and

274 Quenching and Carburising

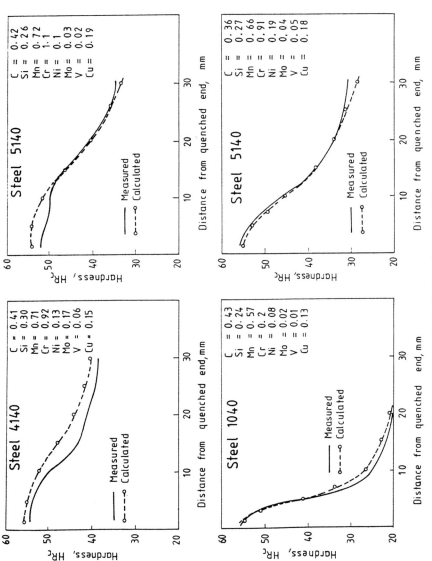

Fig. 3. Comparison of calculated and measured Jominy curves.

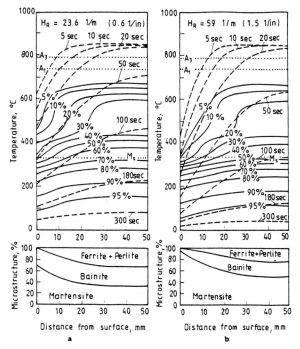

Fig. 4. Transformation diagrams for a 100 mm diameter steel bar with quench intensities of (a) 0.6 1/in. and (b) 1.5 1/in.

the H_R value, a transformation diagram system can be generated from the calculated results for cylindrical or plate-type workpiece and Jominy specimen. Figure 4 shows this type of transformation diagram for two different H_R values (0.6 and 1.5 l/in.) and for the same steel composition and geometry which are indicated in Fig. 2.

In the upper part of Fig. 4 the temperature associated with the percentage of transformed austenite is shown as a function of the distance from the surface of the 100 mm diameter cylindrical part. There are two series of curves in the figure. The dashed curves represent the temperature distribution in the steel part after 5, 10, 20, 50, 100, 180 and 300 seconds; the continuous curves are the loci of points in the workpiece where the quantity of transformed austenite is constant, for values in the range of 5–95%. The lower part of the diagram shows the percentage distribution of the reaction products at room temperature. It is evident from Fig. 4 that the progress of transformation across the cross-section occurs in a non-uniform manner. It can be seen that on cooling the part in a quenchant with $H_R = 1.5$ (l/in.), the temperature at a point of 20 mm from the surface is reduced to 430°C after 40 seconds, when it contains less than 20% of transformed austenite. Figure 4(b) shows that this point contains 55% martensite, 30% bainite, and 15% ferrite + pearlite at room temperature.

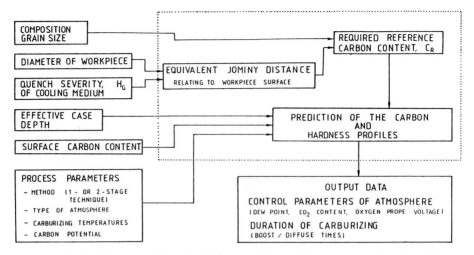

Fig. 5. Block diagram of the simulation model used for process planning of gas carburising.

3 PREDICTION OF PROCESS PARAMETER OF GAS CARBURISING

In what follows, a simulation program is outlined that helps the user in steel selection and the heat treater to minimise the total time required for one- or two-stage gas carburising processes controlled in terms of carbon potential.

The block diagram of the program called PREDCARB is shown in Fig. 5. The computing algorithm simulates the total carburising process, predicts the carbon and hardness profile after quenching and calculates the main process parameters of practical interest.[8] The input data on the left side of the block diagram is as follows:

- Data on the steel grade (chemical composition, grain size).
- Equivalent diameter of the workpiece.
- Quench severity of the quenchant used for quenching after carburising (H_G = Grossman quench factor).
- Effective case depth defined in terms of either a prescribed reference carbon content, C_R, or as a depth to 52.5 HRc required after quenching.
- Data on the process parameters such as technique of carburising, type of atmosphere, carburising temperature in the boost and the diffusion period, carbon potential, and so on.

The output data are:

- Data on the atmosphere composition required to control the carburising process (dew point, CO_2 content, oxygen probe voltage, and so forth).
- Calculated values of the time parameters of carburising.
- Predicted carbon profile and hardness distribution curve generated as a function of the distance below surface.

```
         Input data of the last calculation
            Two stage carburizing          [ 8620H ]

   C   [%]    .2    Si  [%]   .28    Mn  [%]    .8

   Cr  [%]    .5    Ni  [%]   .55    Mo  [%]    .2

 Diameter: 50 [mm]    Cooling intensity of the quenchant: .6  [1/inch]

Type of atmosphere in the boost period: Endothermic gas from natural gas
              CO [%]    20.5          H2 [%]    40.5
Type of atmosphere in the diffuse period: Endothermic gas
              CO [%]    20.5          H2 [%]    40.5
Grain size [ASTM]                                 7
Effective case depth [mm]                         1.25
defined by the 550 HV [52.5 HRc] hardness value
Surface carbon content [%]                        .8
Boost temperature [°C]                            900
Boost carbon potential [%]                        1.15
Diffuse temperature [°C]                          900
```

Fig. 6. Printout of input data for two-stage carburising with endothermic gas based atmosphere.

The algorithm used to predict the hardness distribution on the basis of the calculated carbon profile is based on the method proposed by Wyss.[9] In the following some of the application possibilities of program PREDCARB are demonstrated.

Example 4 (Prediction of process parameters of two-stage gas carburising). Figures 6 and 7 show the printout of input data and the computed process parameters for two-stage carburising of a cylindrical bar made of 8620 H steel, where an endothermic based atmosphere derived from natural gas is used during heat treatment. The predicted carbon and hardness profile displayed at the end of the simulation are given in Fig. 8.

Example 5 (Determination of required carbon profile for a specified case depth). If the effective case depth is defined as depth to a specified hardness such as 52.5 HRc (550 HV), and is used as an input to the model, the reference carbon content C_R necessary to achieve that hardness must be calculated on the basis of Jominy curves. In the case of given steel composition, the reference carbon content C_R depends on the bar diameter D, and the H_R value of the quenchant. This conclusion is demonstrated by computation results for the 4820 steel (0.20% C, 0.65% Mn, 3.5% Ni, 0.25% Mo) in Fig. 9. As can be stated, the required value of C_R may vary significantly as a function of bar diameter and of the cooling power of

278 *Quenching and Carburising*

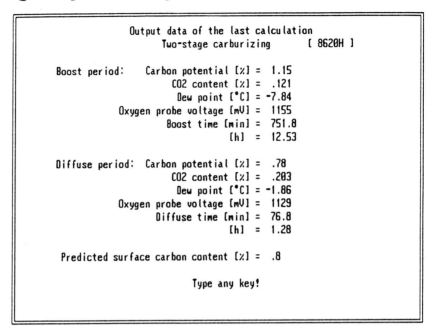

Fig. 7. Printout of predicted process parameters for 2-stage gas carburising.

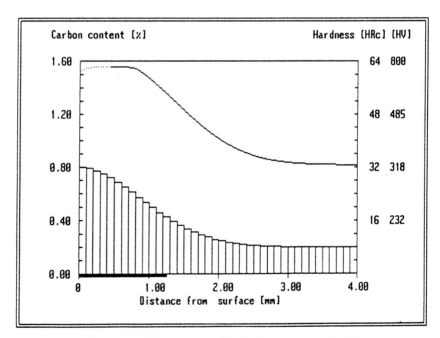

Fig. 8. Calculated carbon and hardness profile for 2-stage gas carburising.

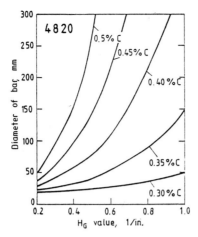

Fig. 9. Relationship between diameter D, quench factor H_G, and the value of C_R necessary to achieve the specified case depth hardness defined as 52.5 HRc (550 HV).

the quenchant. In practice, the value of reference carbon content is chosen in the interval 0.33 to 0.45%. This is because if C_R is less than 0.33, the necessary surface hardness cannot be achieved; on the other hand, if C_R is higher than 0.45%, the required carburising time increases considerably.

Figure 10 shows a comparison between measured and predicted values of effective case depths after boost/diffuse gas carburising and oil quenching.

The cylindrical test specimens of various diameters (30, 50, 75, 100 mm) were

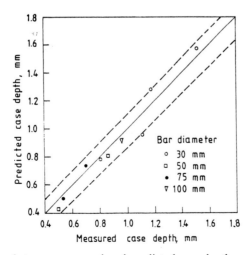

Fig. 10. Correlation between measured and predicted case depths.

machined from low alloy case hardenable steels of different composition. The effective case depth was defined as a depth to 550 HV measured with a load of 500 g. The estimated quench severity was between 0.25 and 0.35. Taking into consideration the accuracy of the hardness test, the agreement is seen to be satisfactory.

CONCLUSIONS

Based on the phenomenological models outlined, computer programs simulating the metallurgical processes occurring during heat treatment have been developed.

The computerised methods can be efficiently used to analyse the influence of different process parameters (steel composition, geometry, section thickness, austenitising and quenching conditions, etc.) on the distribution of microstructure and on the final mechanical properties.

The computer programs implemented on a PC may be utilised to select steel grade using technical and economic considerations and to establish methods of heat treatment that will provide the specified steel properties.

Since the computer program takes into consideration the influence of all relevant factors represented by the input data, the resulting properties can be estimated more precisely. Planning of the heat treating process and preparation of the technological documentation is considerably faster and simpler.

The advantages of the computerised technique are short computing time and precision of computation satisfying the practical requirements. Since the method is simple and user friendly it may be applied readily in plant conditions.

REFERENCES

1. T. Réti, M. Gergely and P. Tardy: *Material Science and Technology, 1987*, **3**, 365–371.
2. M. Gergely and T. Réti: *J. Heat Treating, 1988*, **5**, 125–140.
3. T. Réti, G. Bobok and M. Gergely: 'Computing Method for Non-isothermal Heat Treatment', in *Heat Treatment '81*, The Metals Society, London, 1983, pp. 91–96.
4. B. Liscic and T. Filetin: *Heat Treatment of Metals, 1988*, **5**, 115–124.
5. M. Gergely: *Neue Hütte, 1973*, **18**, 24–26.
6. E. Füredi and M. Gergely: 'A Phenomenological Description of the Austenite-Martensite Transformation in Case-Hardened Steels', *4th International Congress on Heat Treatment of Materials*, Vol 1, West Berlin, 1985, Institut für Werkstofftechnik TU Berlin, pp. 291–301.
7. Sz. Somogyi and M. Gergely: 'Prediction of the Macro-hardness by Help of the Individual Hardnesses of the Microstructural Elements', *4th International Congress on Heat Treatment of Materials*, Vol 1, West Berlin, 1985, Institut für Werkstofftechnik TU Berlin, pp. 84–91.
8. T. Réti, M. Réger and M. Gergely: *Journal of Heat Treating, 1987*, **8**, 55–61.
9. U. Wyss: *Härterei-Technische Mitteilungen*, 1088, **43**, 27–35.

17
Effect of Carburising on Mechanical Properties of Steels

NORIO KANETAKE

Metaltechnic Research Laboratory, 4-5-27 Kamiosaki, Shinagawa-ku, Tokyo 141, Japan

ABSTRACT

The effect of carburising on the mechanical properties of steels is very important in terms of use. This report shows that tensile, bending and torsional strength and toughness of steels are affected by case depth and tempering temperature. It also discusses the behaviour of residual stress and fatigue strength of carburised steels. In the case of carburised chains, tensile stress and bending stress interact in a complex fashion. Breaking stress and ultimate elongation of carburised chains are roughly proportional to each other. Fatigue strength is improved by carburising.

1 INTRODUCTION

The carburising process represents the largest of all productions of surface hardening heat treatment of mechanical parts. Of the effect of carburising on mechanical properties of steels, major factors include case depth, tempering temperature, microstructure of surface layer, surface condition, etc. Of these factors, case depth has the largest effect. For this reason, this paper reports the effect of case depth and tempering temperature on tensile, bending and torsional strength, toughness, and the fatigue strength of Cr-Mo steels (SCM420) and Ni-Cr-Mo steels (SAE8620). It also discusses the behaviour of residual stress of carburised steels.

Complex stress comprising tensile and bending stress occurs in link chains because of tensile force. Chains are thus one of the machine element products that deserves special attention regarding the strength of materials and engineering.[1]

Breaking stress and ultimate elongation of carburised chains are greatly affected by case depth and tempering temperature. The paper also discusses breaking stress, ultimate elongation and fatigue strength.

2 EFFECT OF CARBURISING ON MECHANICAL PROPERTIES

The effect on the mechanical properties (P_M) of steels results from various factors attributable to carburising:

$P_M = f(C_M, F, S, C \cdot D, H, S \cdot L, C \cdot S, \sigma_R)$
C_M = chemical composition of materials
 F = formation factor
 S = condition of surface
$C \cdot D$ = case depth
$S \cdot L$ = condition of surface layer
$C \cdot S$ = microstructure of core
 H = hardness
 σ_R = residual stress

C_M exerts an influence on $C \cdot D$, H, $S \cdot L$ and $C \cdot S$. For S, surface roughness and surface defect affect P_M. $C \cdot D$ (case depth) refers to effective case depth, or $EC \cdot D$. $EC \cdot D$ is defined by 550HV. $EC \cdot D$ is the factor that exerts the greatest influence on mechanical properties. Incidentally, regarding case depth, total case depth, $TE \cdot D$, in addition to effective case depth, also affects the mechanical properties, but in this paper case depth is shown by $EC \cdot D$ assuming that the value $(EC \cdot D)/(TC \cdot D)$ is nearly a constant.

H represents distribution of surface and core hardness and that of the surface layer. $S \cdot L$ is the carbon content, microstructure and intergranular oxidation of the surface layer. $C \cdot S$ is the microstructure, austenite grain size and internal defect of the core. σ_R is the state of residual stress of surface and surface layer.

Hardness, microstructure, residual stress etc. are affected by tempering temperature if the quenching conditions remain constant. This paper describes the relation between, on the one hand, the factors of case depth, hardness, tempering temperature and surface conditions which exert the greatest influence upon mechanical properties and, on the other, the mechanical properties.

3 TENSILE, BENDING AND TORSIONAL STRENGTH

The specimen used was Cr-Mo steel SCM420 (0.23%, 0.72% Mn, 1.12%Cr, 0.21% Mo), with a diameter of 9.5 mm. Heat treatment comprised 930°C gas carburising, followed by oil quenching. Tempering continued for 1 hour at 150–250°C.

Figure 1 shows hardness gradients for the surface layer as the specimen was quenched. Case depth (effective case depth at 550HV) was 0.24, 0.35, 0.53, and 0.68 mm. Carbon content at the surface was 0.80–0.85%C. Figure 2 indicates the effect of case depth on tensile strength. As case depth thickness increases, strength, elongation and reduction of area decrease. Elongation and reduction of area are reduced to zero when case depth becomes approximately 0.5 mm or above. Recommended tempering temperature is 200°C.

Figure 3 shows the result of three-point bending test using a central load of 50 mm span length.

Bending strength, σ_{b-f} (coefficient of bending strength), was obtained in the following equation:

$$\sigma_{b-f} = 8Wl/\pi d^3$$

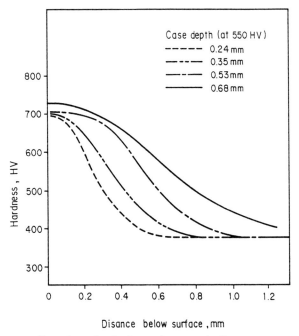

Fig. 1. Hardness gradients in carburised case (as quenched).

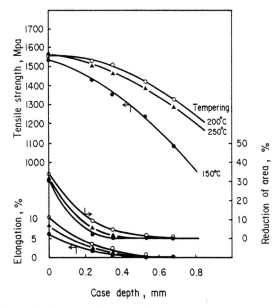

Fig. 2. Effect of case depth on tensile strength.

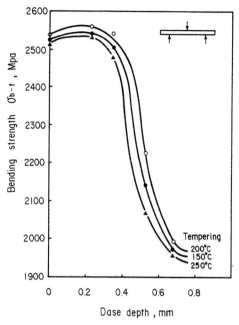

Fig. 3. Effect of case depth on bending strength.

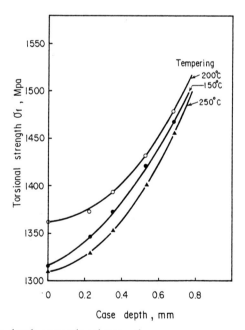

Fig. 4. Effect of case depth on torsional strength.

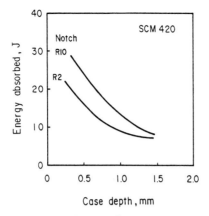

Fig. 5. Effect of case depth on impact energy.[2]

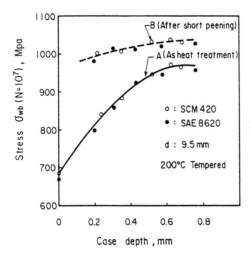

Fig. 6. Effect of case depth on fatigue strength (the method is rotary bending fatigue testing).

where:

W = breaking force,
l = span length,
d = diameter of specimen.

For case depth up to 0.24 mm, $\sigma_{b\text{-}f}$ increases slightly, but above 0.35 mm, $\sigma_{b\text{-}f}$ decreases. The test specimen showed a large number of cracks on the surface because of the load, and was finally broken.

Figure 4 shows the effect of case depth on torsional strength, τ_f. Torsional

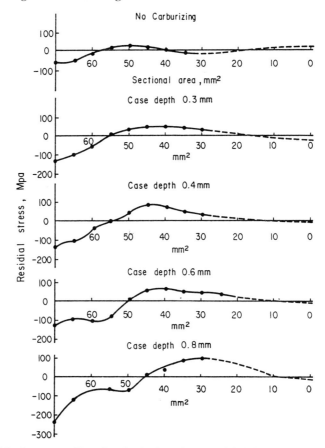

Fig. 7. Residual stress profiles of carburised steels (material: SCM 420, 200°C tempered). (S. Yonetani.)

strength τ_f (coefficient of torsional strength) was obtained in the following equation:

$$\tau_f = 16T/\pi d^3$$

where T = torsional fracture torque.

Torsional strength is remarkably improved as case depth increases. Tensile, bending and torsional strengths all show a high value at 200°C tempering. Brittleness was observed at all other tempering temperatures.

Figure 5[2] shows the influence of case depth on impact value. It is obvious from this figure that the impact value decreases with increasing case depth.

4 FATIGUE STRENGTH OF CARBURISED STEELS

Figure 6 shows the testing of SCM420 (same as the sample tested for tensile strength) and Ni-Cr-Mo steel SAE8620 (0.23%C, 0.75%Mn, 0.62%Ni, 0.56%Cr

Carburising and the Mechanical Properties of Steels 287

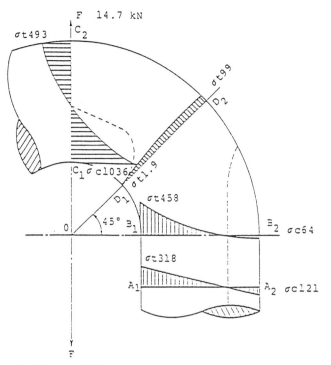

Fig. 8. Stress distribution of link.

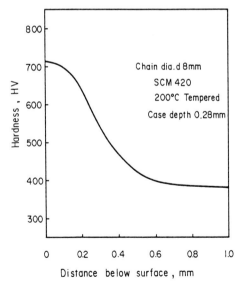

Fig. 9. Hardness gradients in carburised layer of chain.

288 *Quenching and Carburising*

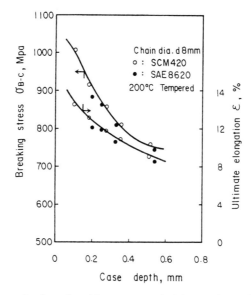

Fig. 10. Effect of case depth on breaking stress and ultimate elongation of chains.

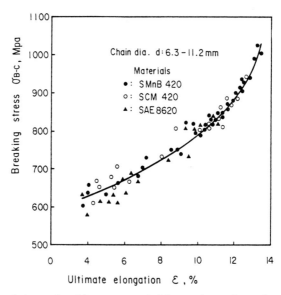

Fig. 11. Relations between breaking stress and ultimate elongations of carburised chains.

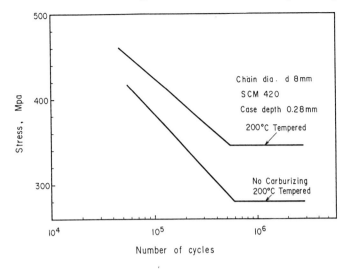

Fig. 12. S-N diagram of carburised chains.

and 0.19%Mo). For a smoothed testpiece of 9.5 mm in diameter, heat treatment conducted comprised 930°C gas carburising and oil quenching, and tempering at 200°C for 1 hour. as shown in Fig. 6, fatigue strength increases with increasing case depth. Fatigue strength increases approximately 40% with carburising of 0.7 mm case depth. In the case of gas carburising, intergranular oxidation exists in the surface layer. To remove it by adding residual stress, shot peening was performed, with the result that fatigue strength was improved compared with case A as shown in Fig. 6. Carburised products are increasingly being treated with shot peening to improve fatigue strength. For carburised products, counter-measures against low cycle fatigue and decreased fatigue strength at long life ($>10^7$) are important.

In addition, vacuum carburising and plasma carburising are increasingly used in industry since they are free from adverse effects of intergranular oxidation. Figure 7^3 shows the effect of case depth on residual stress. The value of compression residual stress on the surface increases with increasing case depth, which corresponds to the test results shown in Fig. 6.

5 MECHANICAL PROPERTIES OF CARBURISED CHAINS

Link chains are widely used in hoists, cranes and conveyors. Chains are carburised for wear resistance and fatigue strength. Carburised chains (case hardened chains) must have strength and toughness, thus optimisation of heat treatment is important.

With chains, tensile strength and bending stress interact in a complex fashion because of tensile force. Thus, chains are one of the mechanical element products needing special attention as regards strength of materials.

Figure 8 shows the stress distribution of chains,[1] Figure 9 is an example of hardness gradients at chain cross-sections and Figure 10 shows the relation between case depth, breaking stress, and ultimate elongation of both SCM420 and SAE8620 chains. Breaking stress and ultimate elongation decrease with increasing case depth.

Figure 11 shows the relation between breaking stress, $\sigma_{B\text{-}C}$, and ultimate elongation ε of chains made of Mn-B steel (SMn420), Cr-Mo steels (SCM4) and Ni-Cr-Mo Steels (SAE8620).

$$\sigma_{B\text{-}C} = 2F_B/\pi d^2$$

where

F_B = breaking force
d = chain diameter

$\sigma_{B\text{-}C}$ and ε are roughly proportional to each other. The relation between breaking stress and ultimate elongation of chains is of two types, 'B-zone,' where they are roughly proportional to each other, and 'T-zone,' where they are reversely proportional to each other.[1]

However, for carburised chains, no T-zone exists for 400°C tempering or below; this applies only to B-zone. B-zone is the brittle fracture region, and T-zone the toughness fracture region. Carburised chains are in the brittle fracture region.

Figure 12 shows an example of a S-N curve of carburised chains. The fatigue strength of chains is improved 24% by carburising. The fatigue strength of carburised chains can be further improved by special treatments after carburising.

CONCLUSION

Various influences can affect the mechanical properties of carburised steels by setting carburising conditions adequately.

(1) Tensile strength, elongation and reduction of area are decreased with increasing case depth. Elongation and reduction of area reduce to zero for case depth approximately 0.5 mm and above.
(2) Even bending strength is decreased by the three-point bending test with increasing case depth. In the bending test, cracks occurred on the surface of the specimen, causing the specimen to break immediately.
(3) Torsional strength is improved as case depth is increased.
(4) Fatigue strength is improved approximately 40% by the carburising of 0.7 mm case depth.
(5) Fatigue strength can be further improved if shot peening is performed after carburising. The effect of shot peening is the exclusion of intergranular oxidation on the surface and the addition of compression residual stress.
(6) For carburised chains (case hardened chains), breaking stress and ultimate elongation are decreased with increasing case depth.
(7) The relation between breaking stress and ultimate elongation of carburised

chains is roughly in proportion, being in relation to the B-zone of the brittle fracture region.
(8) Fatigue strength and wear resistance of chains can be improved by carburising.

REFERENCES

1. N. Kanetake: *Proceedings of the 7th International Congress on Heat Treatment of Materials (Moscow)*, Vol. II (1990), p. 22.
2. T. Inaba et al.: *Proceeding of the Symposium of Mechanical Properties of Case Hardened Steels*, Japan Society for Heat treatment, (1984), p. 8.
3. According to S. Yonetani's data.

18
Heat Treatment with Industrial Gases – Linde Carbocat® and Carboquick® Processes

ABUL KAWSER

Gas Applications Department, Linde Gas Pty Ltd, Fairfield, NSW, Australia

ABSTRACT

Shielding gases and mixtures of shielding and reactive gases are being employed in heat treatment process to produce or prevent specific changes in the materials. In a large number of annealing processes, for example, expensive after treatment can be avoided through selection of the correct furnace atmosphere. Processes such as gas carburising or sintering would not be possible at all without the use of gas mixtures.

Linde Carbocat® is a process for catalytic production of a reactive protective gas within a heat treatment furnace. Carbocat® offers an alternative to the employment of exothermic gas, nitrogen methanol or nitrogen-natural gas. Carbocat® can thus be employed to those heat treatment processes in which exothermic gas is being used e.g. bright annealing and recrystallisation or stress-relief annealing of semi-processed pieces.

Carbocat® represents a significant advance in terms of cost efficiency and environmental compatability. All existing and new heating tube continuous and pusher furnaces are suitable for the employment of Carbocat®, as well as isothermal annealing plants and decarburising furnaces. It operates at furnace temperatures between 600°C and 1000°C without any problems. It produces a higher quality reactive shielding gas more economically than exothermic gas produced in generators.

The Linde Carboquick® process is used in conjunction with the Carbocat® process to provide higher cooling capacity during quenching operation, enabling higher furnace throughput. The concept of turbulent flow for higher heat transfer capacity as compared to laminar flow is used in the design of the Carboquick® process.

1 INTRODUCTION

Due to advances in gas technology, the majority of heat-treatment processes for ferrous and non-ferrous metals are carried out using synthetic gas atmospheres. This applies especially to those processes using reactive gases (e.g. quench hardening) where the use of synthetic atmosphere lowers production costs.

The Carbocat® process, developed by the Linde Applications team, provides the right type of synthetic gases as protective furnace atmosphere, for heat treatment purposes replacing the generator gas system. The Carboquick® process, used together with Carbocat® represents the perfect furnace gas supply system. This paper discusses in detail these two processes and their applications including recent developments.

1.1 THE CONVENTIONAL DX GENERATORS

The shielding gas requirement of roller hearth and other continuous furnaces for normalisation, recrystallisation, stress-relieving and isothermal annealing of semifinished products and steel castings is currently being met primarily by DX (exothermic) gas generators.

The shielding gas thus produced, although economical given the current cost of natural gas, nevertheless has a number of considerable disadvantages. Firstly, exogas exerts a substantial decarburisation effect on the product. Secondly, problems can occur with regard to the emission of pollutant. Moreover, the exogas generators employed require a great deal of maintenance. Unattended weekend furnace operations involving the use of shielding gas thus involve a considerable amount of risk. So much so, in fact, that this profitable mode of operation has to be discounted completely.

1.2 INCREASED APPLICATION OF LIQUID SUPPLY GASES

For the above reasons and owing to a growing demand for carbon-neutral annealed semi-finished products, many cold rolling mills, wire drawing plants, tube manufacturers and other producers of semi-finished products in the steel and non-ferrous metal industries have, over the last few years, shut down their gas generators and replaced them with tank-supplied process gases.

With the increased employment of liquid supply gases by manufacturers of semi-finished products, gas producers have been actively cooperating with their users in the development of new protective gas processes which increases the cost-efficiency of the heat-treatment plant and improve product quality.

The simplest alternative to DX gas from the point of view of the installation is to employ tank supplied nitrogen with an addition of 2–3% natural gas as the reducing agent. The savings in shielding gas costs which can be achieved are, however, still not high enough in order to enable them to compete effectively with the DX gas method. In addition, there is the possibility of problems occurring with carbon deposits in cold furnace zones.

A further possible variant involves the employment of nitrogen as the carrier gas with the additions from a liquid methanol supply. This method (the Carbothan® process) is widely used for case-hardening and tempering processes in heat-treatment plant. It can however only be employed for annealing semi-finished steel products if the treatment temperatures are above 800°C. Below this furnace temperature, methanol cracking into the gas components CO and H_2 remains incomplete. As a result, undesirable by-products such as soot, CO_2 and aldehyde are formed which can be corrosive as well as causing unpleasant odours outside the furnace. In such cases, the methanol has to be precracked – but that means the DX generator is merely replaced by another type of generator.

For heat treatment facilities which also have to be operated at temperatures below 800°C, use of methanol does not constitute a viable solution, particularly in view of the additional investment required for the methanol tank facility, handling station, pipeline reticulation system and control equipment.

1.3 THE AIM OF CARBOCAT®

The aim of the development of the Carbocat® method was to achieve the advantages of the nitrogen/liquid methanol system with regard to high quality of annealing but to avoid the problems associated with this system.

The first objective was to avoid the need for additional media apart from nitrogen as the inert carrier gas, other than those already available at the furnace for the generation of shielding gas.

In addition, the aim was to produce a gas which was highly reducing, which has high thermal conductivity and contained only small amounts of oxidants such as CO_2 and H_2O and which was cleaner and less inflammable than the exothermic gas conventionally used. Furthermore, it was important that the supply of shielding gas should save costs.

1.4 THE CARBOCAT® PRINCIPLES

It was clear that all these characteristics could not be combined by a special shielding gas, but only by means of a special sparging system.

For the generation of the reactive gas, the obvious method was the use of the cost-effective natural gas and atmospheric air which were both available at the furnace site.

The intention was to generate a reactive gas according to the equation:

$$2CH_4 + O_2 + 3.8\ N_2 \xrightarrow{950°C} 2CO + 4\ H_2 + 3.8\ N_2 \quad (1)$$

which contained approximately:

CO 20%
H_2 40%
N_2 39%

and a small amount of CO_2 and H_2O. In other words, an endothermic gas which is otherwise used chiefly for gas carburising.

A major benefit of this system in comparison to the use of methanol is that almost 3.8 mol. of nitrogen is gained from air, at almost no cost.

This can be used for overall supply of shielding gas and does not have to be taken from a tank. With a natural gas/air ratio of approximately 1/2.4, almost 5 m³ of

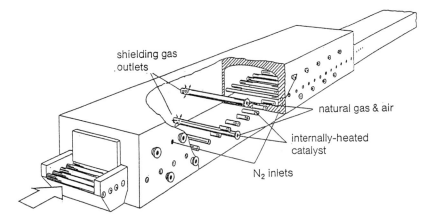

Fig. 1. Continuous annealing furnace using Linde-Carbocat®. Production of the reactive gas is achieved by catalytic combustion of a natural gas/air (or propane/air) mixture in a radiation tube converted to a catalyst in the furnace.

shielding gas can be generated from 1 m³ of natural gas. A litre of methanol, by contrast, only produces 1.67 m³ of cracked gas.

2 DESCRIPTION OF THE CARBOCAT® PROCESS

In the Carbocat® process, one or two radiant tubes per furnace are converted into heated protective gas retort filled with a catalyst. In this radiant tube, a CO-, H_2-, and N_2-containing reducing gas, which has a small CO_2 content, is produced from a natural gas/air mixture. This highly reactive reducing gas is transferred from the retort directly to the furnace chamber in its hot condition. The catalytic conversion requires a certain minimum temperature. In order to achieve and maintain the temperature for the endothermic process, heat is taken directly from the furnace, or, if this is not sufficient, an internal burner will be operated in the catalyst tube. The separate heating of the catalyst ensures that the requisite conversion energy is always available even in the case of fluctuating or reduced furnace temperatures. This means that even at annealing temperatures under 800°C, shielding gas of a good quality is always present in the furnace. Figure 1 shows a schematic diagram of a roller hearth furnace employing the Carbocat® process.

With the Carbocat® process, a high level of reactivity of the protective gas is maintained in the heating zone. As a result of the zonal gas supply pattern produced, the furnace atmosphere can also be controlled so that substantially higher CO and H_2 contents are present in the hot reaction zone than would be achievable with a DX gas. The gas entering and leaving the furnace is, however, diluted by the nitrogen infeed to such an extent that its combustibility values and pollutant contents are below those of DX gas.

Heat Treatment with Industrial Gases 297

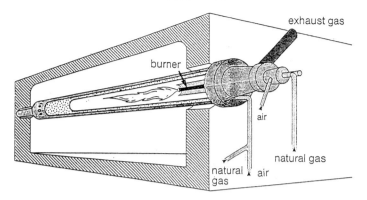

Fig. 2. Construction and function of Carbocat® heating tube.

Figure 2 shows the construction of a catalyst tube. The catalyst gases, which are of similar composition to endothermic generator gases, are fed to the highest temperature and thus most reactive zone of the furnace, so that an efficient reaction with the fed material is ensured. Nitrogen supplied to the furnace (independently to the catalyst gases) at points before and after the catalyst tube provides blanketing effect, sealing the furnace against the ingress of any air and producing a reaction zone.

The significantly lower CO_2 content compared with exothermic gases means a 30–50% decrease in the volume of shielding gas required. At the same time, the Carbocat shielding gas has a higher carbon-potential which reduces decarburisation effects at higher temperatures.

The protective gas is carbon-neutral, although additional hydrocarbon enrichment will produce carburising conditions. As a result of the high H_2 content in the heating zone, the conditions are more conductive to efficient heat transfer to the product.

The flexibility of the Carbocat® process means that production of catalytic gas can be stopped at any time, i.e. during shut-down and weekend operations, and the furnace supplied with nitrogen (at a lower flow rate) to maintain a neutral furnace atmosphere.

2.1 ADVANTAGES OF THE CARBOCAT® PROCESS

From the discussion above we can summarise the following benefits of this process:

- elimination of the gas generators susceptible to failure;
- possibility of carbon-neutral annealing;
- more economical energy utilisation;
- possibility of unattended weekend operations;
- reduced waste gas emissions.

2.2 CARBOCAT EQUIPMENT

2.2.1 Gas supply

For operation and heating of the catalyst, air and natural gas must be available at the furnace. A certain minimum pressure must be maintained for both substances. Nitrogen is stored in liquid nitrogen tanks and after passing through vaporisers is pressure fed via a pipeline to the furnace.

2.2.2 Control units

The Carbocat® process demands exact and therefore sophisticated control of the natural gas/air ratio. Thus each catalyst requires a control unit, which has been developed by Linde.

The control units perform the following functions:

- measurement and control of natural gas/air supply to the catalyst;
- measurement and control of natural gas/air supply to the internal burner of the catalyst;
- electronic control of the natural gas/air flow with temperature control and safety switching;
- alarm, on any faults/failures;
- measurement and control of nitrogen supply to the furnace.

2.2.3 Carbocat® – Applications

In the steel and non-ferrous industry, Carbocat® has the following applications

In: cold rolling factories, tube factories, drop-forge factories, foundries.
For: continuous annealing furnace, pusher furnaces, rocker-bar furnaces, chain conveyor furnaces.
To: crystallisation annealing, stress-relief annealing, bright annealing, isothermal annealing.
Of: strip steels, tin coils, wire coils, tube coils, sectional steel, tube, beamless and welded, forged pieces, cast pieces.

2.2.4 Carbocat installation

One or more catalyst tubes may be used depending on the size of the furnace. The catalyst tube and burner are fitted within one radiation tube, which has the advantage that with a gas-fired furnace the catalyst unit may be installed in the existing radiation tubes without any alterations to the furnace itself or a break in production. The alteration of a radiation tube to a Carbocat® catalyst is relatively easy.

Feeding of nitrogen can be done through existing openings, e.g. inspection ports. Maintenance requirements for the Carbocat® units are generally very low.

Figure 3 shows a continuous furnace equipped with the Carbocat® protective gas system. Operators check the temperature values in the radiant tube. Figure 4

Fig. 3. Operators check the temperature value in the radiant tube in a continuous furnace equpped with Carbocat® facility.

shows a roller hearth furnace for annealing semi-finished products with the Carbocat® facility.

3 OPERATIONAL DATA

The following operational data were recorded for the Carbocat® process employed in a continuous furnace:

Heating tube without a burner:
  Temperature in catalyst bed
    950°C
  Natural gas
    3.5 m^3 h^{-1}
  Air
    9.5 m^3 h^{-1}
  Gas compositions
    $CO = 15-20\%$,
    $CO_2 = 0.1-0.6\%$,
    $H_2 = 30-32\%$,
    Balance N_2.

Heating tube with a burner:
  Temperature in catalyst bed
    900–1022°C

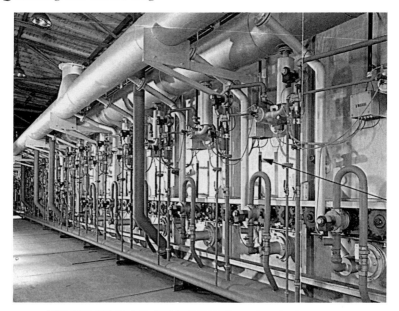

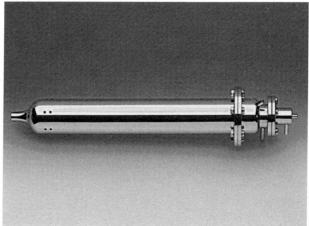

Fig. 4. Roller hearth furnace for annealing semi-finished products with Carbocat® facility

Natural gas
  3–4 m³ h⁻¹
Air
  7.2–9.5 m³ h⁻¹
Gas compositions
  CO = 19%,
  CO_2 = 1.7%,
  H_2 = 31%,
  Balance N_2.

Gas flow to the burner: natural gas 1 m³ h⁻¹. Air 20 m³ h⁻¹.

A typical measurement of atmospheres at different zones of a roller hearth furnace are obtained when following gas flows are employed:

Gas flow to catalyst bed:
  3 m³ h⁻¹ natural gas + 7.2 m³ h⁻¹ air.
Gas flow to burner:
  1 m³ h⁻¹ natural gas + 20 m³ h⁻¹ air.
Nitrogen flow to inlet and outlet of heat zone:
  50 m³ h⁻¹.

The reactive gas composition at the heating zone between nitrogen inlet and outlet is:

  CO 4.2%
  H_2 8.5%
  CO_2 0.83%
  CH_4 0.4%
  T_p +13°C

The gas composition before nitrogen inlet point to the furnace is:

  CO 3.5%
  H_2 5.4%
  CO_2 1.23%

3.1 GOOD RESULTS

So far, Linde Gas in Germany has installed 25 Carbocat® facilities, predominantly in plants manufacturing tubing and steel bar. The results achieved have been excellent. Moreover, facilities of this kind are also installed in pusher-type furnaces employed in the mechanical engineering sector and in isothermal annealing plant in the automative industry.

However, because the process only becomes economically viable in the case of heat-treatment facilities with a shielding gas consumption level of more than 40 m³ h⁻¹, Linde has developed a furnace networking system. In this concept, a furnace equipped with a Carbocat® facility also functions as the shielding generator for a number of other furnaces. Thus, given an appropriate furnace arrangement, the Carbocat® system can also be employed economically for smaller individual heat-treatment units.

4 RECENT DEVELOPMENT IN THE CARBOCAT® PROCESS

A new Carbocat heating tube has been developed by Linde Gas to overcome the problem associated with low furnace temperature. At a furnace temperature of around 700°C, the temperature inside the heating tube could be less than 800°C, resulting in incomplete conversion of natural gas and air to its reducing components. As a result soot formation was evident in some cases. The newly developed heating tube has an outer tube, with slots across the length of the tube. The old heating tube is inserted into this tube.

The hot reactive gas mixtures coming out at the end of the old tube are retained in the outer tube before being expelled through the slots. The reheating of catalyst with the hot gas produces higher temperature inside the catalyst and maintains uniformity of temperature all around. The reinsertable catalyst heating tube also provided flexibility in the maintenance of the heating tube.

Another development is in the design of a new catalyst. The catalysts used in the heating tube are ceramic balls (10 mm diameter) coated with nickel. This type of catalyst had a relatively shorter life due to the fact that during a rich natural gas/air mixture soot formed in the tube deposits in the porous surface of the catalyst, while during a rich air/natural gas mixture soot burns and generates heat, causing distortion of the catalyst, rendering it ineffective. The development of a less porous carrier metal for nickel which is thermo-chemically more stable has increased the catalyst life significantly. The Linde Gas application team is currently investigating the possibility of using palladium as the catalyst in the heating tube.

In the Carbocat® process, normally a gas mixture of natural gas and air in the ratio of 1:2.5 is normally used at temperature above 645°C. However, an experiment is being conducted using less amount of propane (ratio propane:air 1 to 6) in air at temperature around 510°C.

The following compositions were obtained in the heating tube:

CO 23%
H_2 31%
N_2 46%

The use of propane has the following benefits:

(1) Higher nitrogen production – more economic, thus less supplied nitrogen is required.
(2) It has more carburising potential.
(3) Cracking of gas at a lower temperature.
(4) A mixture of a small amount of propane in natural gas and air can produce a condition similar to carburising.
(5) Sulphur, often added to the natural gas, can form nickel sulphites, making the nickel base catalyst ineffective. If the sulphur level in natural gas is above 6 mg m^{-3}, it is advisable to use propane or a mixture of propane and natural gas in the heating tube.

5 DESCRIPTION OF THE LINDE CARBOQUICK® PROCESS

In many annealing shops, the cooling capacity of the cool down zone in continuous furnaces represents the bottleneck that prevents throughput from being increased.

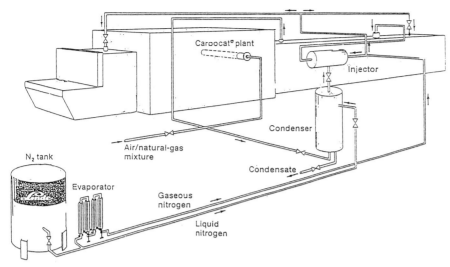

Fig. 5. Carboquick® and Carbocat® process.

Frequently, it is not possible to extend the length of the cooling zone due to reasons of space or cost. The Linde Carboquick® process has been developed to overcome this problem.

Figure 5 shows the Carboquick® process in conjunction with the Carbocat® process. Together they represent the perfect furnace gas supply system. Under this process, shielding gas is extracted at the beginning of the cooling zone, advanced through the watercooled heat exchanger, further cooled against liquid nitrogen in an alternating condenser, dried, and then returned to the cooling zone in the form of cold gas. The gas is advanced by a vapour pump that operates with vaporised nitrogen. It doubles as both a pump and a gas mixer, without containing any mechanically driven components. The cold pressurised vapour, introduced in the cooling zone in a counter flow pattern to the parts, produces a turbulent flow enabling an increase in the heat removal capacity.

Typical advantages of the Carboquick process are:

- Higher cooling capacity, around 25%.
- No misting at the furnace outlet.
- Additional drying of the shielding gas, thus resulting in higher purity of the shielding gas in the cooling zone.
- No expensive controls.
- No distortion of the annealed product.
- No temperature drop below the dew point in the cooling zone.

6 CONCLUSION

Linde Carbocat® and Carboquick® processes together provide the perfect solution for heat treatment with the supplied gas system. From an economic and environmental point of view Carbocat® is an effective alternative to exothermic and other similar gases. Compared with generator gases, savings in shielding gas requirements of 30–50% have been achieved with Carbocat®. Compared with nitrogen/methanol and nitrogen/natural gas, savings of 20–30% have been observed. Carbocat® also enables annealing at a temperature between 600°C and 1000°C. The Carbocat® process has been thoroughly proven as an ideal solution for a wide range of applications, a range which is likely to expand even further in the future.

Carboquick® increases the cooling efficiency of a furnace during quenching by 20–25%. An increase in throughput together with high quality of quench has made Carboquick® a very effective heat treatment option.

Index

air–mist nozzles, for spray quenching 161–75
  and air–water ratios 173–4
  distribution measurements 164–6
  droplet size tests 162–4
  experiments on 162–71
  results 171–4
  steady-state test 168–71
  unsteady-state test 166–8
argon gas quenching in vacuum furnaces 111–12
  with helium mixture 114–15
austempering
  in fluidised bed quenching 103
  of steel 255–8
    applications for 260
    influence on component properties 258–60
    nodular cast iron 260–2
    plants for 262–4
austenite
  microstructure of 207–10
  transformation, kinetics of 130–2

bending strength, and mechanical properties of steel 282–6
boiling phase in cooling 35, 178

Carbocat®
  advantages 297
  aim of 295
  equipment 298–9
  good results 301
  operational data 299–301
  principles of 295–6
  process description 296–9
  recent developments 301–2
carbon
  diffusion 227
    in steels 233
  equilibrium gas composition 230–3
Carboquick process 302–3
carburised components
  fatigue strength of 286–9
  microstructures in 235–7
  ductility of 241–7
  toughness of 237–41
toughness of 227–49
carburising
  austenite 207–10
  carbon diffusion in steels 233
  chains, mechanical properties of 289–90
  equilibrium gas composition 230–3
  fatigue in *see* fatigue
  fundamentals of 229–35
  kinetics of 233–5
  martensite–austenite–carbide 211–12
  and mechanical properties of steel 281–91
  see also gas carburising
case hardened steels, property prediction 267–80
Castrol/Renault hardening power 48–52
CCT *see* continuous cooling transformation
chains, carburised, mechanical properties of 289–90
continuous cooling transformation (CCT)
  austenite transformation, kinetics of 130–2
  diagrams
    and cooling curves 43
    and cooling rates, comparison of 103–4
  physical properties in 128–9
  quenching heat transfer coefficient 129–30
  and residual stress 128–32
controlled atmosphere plants for martempering of steel 253–5
convection
  forced, in gas quenching 109–11
  phase, in cooling 35, 178
cooling
  characteristics of 177–8
  performance, and cooling curves 52
  rates, comparison of 104
  simulation of in-line heat treatment of steel 189–203
    applications 199–202
    continuous cooling 195–7
    cooling system 191–3
    modelling results 197–9
    multistage cooling 202
    nozzle performance 195
    rig design 190–5

306 *Index*

cooling (*contd*)
  simulation etc (*contd*)
    spray nozzles 193
    temperature measurement 193–5
  of steel spheres
    first-stage 122–3
    second-stage 123–4
  and temperature intervals 44
  water, heat transfer in 146–51
  see also continuous cooling
cooling curves
  and CCT diagrams 43
  and cooling performance 52
  evaluation of quenching intensity by 4–8
  and hardening, correlations 44–52
  in testing quenching power 34–9
cooling power
  cooling curves in 34–9
  hot wire test of 40
  Houghton quench test of 39–40
  interval test of 41
  magnetic test of 40
  testing, methods of 34–41
  thermal gradient test of 39

deformation models in thermomechanical treatment of steel 140–3
  post-deformation cooling models 143–57
dragout losses, reducing 74–5
droplet size tests in spray quenching 162–4
  and air–water ratios 173–4
  distribution measurements 164–6

environmental hazards caused by quenching media 71–2

fatigue in carburised steels 205–25
  fatigue mechanisms and performance 221–4
  and residual stress
    general considerations 212–14
    measurement and modelling 214–15
    quenching and tempering effects 216–19
    and refrigeration 219
    and shot peening 215–16
    strain-induced austenite transformation 219–21
    and surface oxidation 215
film quenching 161
fluidisation velocity in fluidised beds 90–2
fluidised bed quenching 85–105
  additives to 93
  austempering 103

fluidisation velocity 90–2
and gas quenching 104
marquenching tool steels 98–103
and oils quenching 97–8
part configuration and densely packed loads 93–5
particle size 86
particle type and density 86–90
and salt-bath quenching 98–103
supporting gas composition 92–3
theory of 85–95

gas carburising 206–7, 229
  prediction of process parameter 276–80
gas composition in fluidised beds 92–3
gas quenching
  cooling, modelling of 109–13
  experimental results 114–16
  and fluidised bed quenching 104
  with gas mixtures 112–14
  with helium 107–18
    with higher velocity 116
    with pressure variation 115–16
  helum-argon mixture 114–15
  with industrial gases 293–304
  operation of 108
  parameters controlling cooling 108–9
  principles 108–13
  in vacuum furnaces 21–4
gases
  industrial, heat treatment with 293–304
    and Carbocat *see* Carbocat®
    Carboquick® process 302–3
    conventional DX generators 294–6
    liquid gases, increased supply of 294–5
  liquid, increased supply of 294–5

hardening
  and cooling curves, correlations 44–52
  evaluation of quenching power 33–54
hardening power of quenching media
  immersion quench tests 41–2
  Jominy end-of-quench tests 42–3
  testing 41–3
heat flux density, evaluation of quenching intensity by 8–13
heat transfer
  air-mist nozzles, for spray quenching 161–75
  conditions, in laboratory and in industry 177–88
  steady-state test 168–71
  unsteady-state test 166–8

in water cooling in thermomechanical
treatment of steel 146–51
heat transfer coefficient
batch, effect on cooling 184–7
temperature/time curves in 186–7
test procedure 184–6
calculation procedure for 179
in continuous cooling 129–30
for cylindrical test probes 180
test procedure for 178–9
variation around a part 180–4
calculated 183–4
calculation procedure 181–3
test procedure 181
heat treatment
fluidised beds for quenching 85–105
with industrial gases 293–304
helium gas quenching in vacuum furnaces
107–18
with argon mixture 114–15
with higher velocity 116
with pressure variation 115–16
hot wire test of cooling power 40
Houghton quench test of cooling power
39–40
hydrogen gas quenching in vacuum furnaces
111–12

immersion quenching 3–4, 161
and tests of hardening power 41–2
in-line heat treatment of steel,
simulation of 189–203
applications 199–202
continuous cooling 195–7
cooling system 191–3
modelling results 197–9
multistage cooling 202
nozzle performance 195
rig design 190–5
spray nozzles 193
temperature measurement 193–5
intensive quenching 27–30
interval test of cooling power 41
IVF hardening power 45–8

Jominy end-of-quench tests 42–3

magnetic test of cooling power 40
marquenching salts as quenching media 81–2
consumption of, reduction of 82
and tool steels 98–103
martempering of steel 251–5

in salt baths 251–3
with controlled atmosphere plants
253–5
martensite-austenite-carbide microstructure
211–12
mechanical properties of steel, and
carburising 281–91
microstructure
austenite 207–10
in carburised components 235–7
ductility of 241–7
toughness of 237–41
and carburised steel, fatigue in 205–25
evolution model in thermomechanical
treatment of steel 141–3
martensite-austenite-carbide 211–12
and processing steels 206–12
model for simulating quenching 132–6
output 136
principles 132–3
procedures 133–6

nitrogen gas quenching in vacuum furnaces
111–12

oils as quenching media 72–6
dragout losses, reducing 74–5
and fluidised bed quenching 97–8
minimising consumption 73–4
polymers, substituting for 77–8
vapours from, hazards of 75–6

pack carburising 229
particles in fluidised beds for quenching
size of 86
type and density 86–90
phase transformation in model for
simulating quenching 133–4
plasma carburising 229
polymers as quenching media 76–81
extending life of 78–80
substituting for oils 77–8
waste disposal of 80–1

quench factor analysis 13–19
quench factor (Q) 58
calculation 58–61
concepts 57–68
development of 57–8
and modelling hardness data 68
and quenchant characterisation 61–8

Index

quenched steels: property prediction 267–80
  after quenching and tempering 268–76
  applications 272–6
  and gas carburising 276–80
  model description 268–72
quenching
  conditions, in laboratory and in industry 177–88
  intensity, evaluation of
    by cooling curves 4–8
    by heat flux density 8–13
  model for simulating 132–6
  prospects for 30–1
  residual stress during 127–38
  and self tempering processing (QST) 189, 199–202
  severity, and steel properties 55–70
  state of art in 1–32
  and steel properties 43–52
  and tempering effects on residual stress 216–19
  see also fluidised bed; gas quenching; immersion quenching; spray quenching
quenching heat transfer coefficient in continuous cooling 129–30
quenching media 1–2
  comparisons between 95–104
  cooling rates, comparison of 104
  hazards caused by 71–2
  marquenching salts 81–2
  oils 72–6
  polymers 76–81
  quenching power, evaluation of 33–54
  testing quenching power, methods of 34–41
  use and disposal of 71–83
quenching power
  evaluation of 33–54
  testing, methods of 34–41

residual stress
  computer simulation of 127–38
  fatigue mechanisms and performance 221–4
  and fatigue of carburised steel 212–21
    general considerations 212–14
    measurement and modelling 214–15
    quenching and tempering effects 216–19
    and refrigeration 219
    and shot peening 215–16
    strain-induced austenite transformation 219–21
    and surface oxidation 215
  in steel spheres 119–26

salt baths
  carburising 229
  martempering of steel 251–3
    with controlled atmosphere plants 253–5
  quenching
    and fluidised bed quenching 98–103
    new developments in 24–7
shot peening of steel and residual stress 215–16
spray quenching 19–21
  air-mist nozzles for *see* air-mist
  simulation of in-line heat treatment of steel *see* in-line heat treatment of steel
steady-state heat transfer test 168–71
  comparisons with unsteady-state 171–3
steel
  austempering of 255–8
    applications for 260
    influence on component properties 258–60
    nodular cast iron 260–2
    plants for 262–4
  carburised
    fatigue strength of 286–9
    mechanical properties of 281–91
  carburised, fatigue in 205–25
  case hardened, property prediction 267–80
  martempering of 251–5
  properties of
    and quench severity 55–70
    and quenching performance 43–52
  quenched, property prediction *see* quenched steel
  spheres, quenching of 121
    calculation using model 122–4
    primitive model of 121–2
    results 124–5
  thermomechanical processing 139–59
strain
  energy minimisation in model for simulating quenching 135
  thermal, generation of 120–1
strain-induced austenite transformation, and residual stress 219–21
stress
  calculation in model for simulating quenching 134–5
  thermal, generation of 120–1
  surface oxidation of steel and residual stress 215

Tamura *V* value, in testing hardening power 44–5

temperature/time curves in batch cooling 186–7
tempering
  effects on residual stress 216–19
  property prediction after quenching 268–76
tensile strength, and mechanical properties of steel 282–6
thermal gradients in testing quenching power 34–9
thermal model in thermomechanical treatment of steel 140–1
thermomechanical controlled processing of steel (TMCP) 139
  spray cooling unit for 189–203
thermomechanical treatment of steel (TMT) 139–59
  deformation models 140–3
  heat transfer in water cooling 146–51
  microstructure evolution model 141–3
    ferrite grain size prediction 156–7
    transformation kinetics 151–6
  post-deformation cooling models 143–57
  spray cooling unit for 189–203
  structure property relationships 157
  thermal model 140–1

torsional strength, and mechanical properties of steel 282–6
transformation kinetics
  austenite 130–2
  thermomechanical treatment of steel (TMT) 151–6

unsteady-state heat transfer test 166–8
  comparisons with steady-state 171–3

vacuum furnaces, gas quenching in 21–4
  with helium 107–18
vapour film phase in cooling 35, 178

waste disposal of polymer quenchants 80–1
water
  and air ratio in droplet size tests 173–4
  heat transfer in cooling 146–51

yielding in model for simulating quenching 135